FRANCIS BACON
BARON OF VERULAM, VISCOUNT ST. ALBANS

The New Organon
and Related Writings

Edited, with an Introduction, by
FULTON H. ANDERSON
Professor of Philosophy, University of Toronto

The Library of Liberal Arts
published by
THE BOBBS-MERRILL COMPANY, INC.
INDIANAPOLIS · NEW YORK

CONTENTS

INTRODUCTION

I. *Some Biographical Data*

Francis Bacon was born in 1561 and died in 1626. He was thus a contemporary of Galileo, almost a hundred years the junior of Copernicus, and thirty-five years older than Descartes. He was one—Descartes the other—of the two original philosophers produced in that period, germinal of science, rich in the arts, and abundant in eclectic thought, called the Renaissance. His days overlapped those of Drake, Raleigh, Spenser, Shakespeare, and he evinced the same imaginative daring which characterized these other "great Elizabethans." Drake voyaged around the world, adventuring in strange seas, and he and Raleigh stormed the strongholds of the Spanish Main, while Bacon embarked on bold intellectual enterprises, assaulted the fortresses of learning, and in a philosophic way encompassed the whole realm of nature. Like Shakespeare in his plays, Bacon in his essays, histories, and philosophical writings probed the passions, thoughts, and faiths which motivate the actions and determine the opinions both of great and of little men.

Through the circumstances surrounding his family, and because of his plans for the support of his philosophical designs, Bacon throughout his life was involved in the affairs of royal courts of worldly magnificence and very considerable learning. He was the son of Sir Nicholas Bacon—Lord Keeper of the Great Seal and senior legal officer in the Kingdom, the nephew by marriage of Lord Burghley—Elizabeth's Lord Treasurer and her most trusted adviser, and the cousin of Lord Salisbury, Burghley's son—Secretary of State and Lord Treasurer under James I. Bacon's mother, Anne Cooke, was a sister of Lady Burghley and a daughter of Sir Anthony Cooke, tutor to Edward VI. Lady Bacon, herself schooled in Greek, Latin, French, and Italian, was the demanding and

commanding person who subjected her gifted son to a rigor-
ous discipline in ancient and modern authors. In religion a
Calvinistic Puritan with "fanatical" tendencies, Lady Bacon
did not hesitate to upbraid her brother-in-law, Burghley, for
his part in Elizabeth's repressive measures against "noncon-
formists" both within and without the Established Church.
To her early influence may be credited her son's segregating,
in Calvinistic fashion, the principles of philosophy from the
dogmas of religion given through revelation, and his pleas,
in unheeded "advices," to Elizabeth and James for a liberal
ecclesiastical policy in state affairs.

Francis Bacon entered Cambridge in 1573, when twelve
years of age. At the University he read Aristotle and his Peripatetic commentators and some of Plato and his Augustinian
interpreters. University exercises for the training and the
examining of candidates consisted mainly of disputations.
These were conducted according to the rules of syllogistic
logic. A respondent was required to defend theses, with terms
defined, against two or more opponents. The candidate's first
disputations were rhetorical exercises, considered preparatory
to later disputational "demonstrations of truth." "Truth" in
this conjunction consisted largely of a collection of propositions traditionally taken from the physical, ethical, political,
and metaphysical works of Aristotle. The contexts of these
propositions in the original writings were more often than not
unfamiliar to candidates and not always remembered, if ever
known, by presiding officers. Against this method of testing
and examination Bacon rebelled. To him it was no more
than verbal gyration, elevated and refined by the Peripatetics into an art. A Peripatetic, it seemed, having assumed
some "principles," could through the art of logic compose a
complete system of discourse, like a playwright making a play
with little, if any, foundation in fact. The Peripatetic method
of proof was a bequest from the "magisterial" Aristotle, who
had always, in Bacon's opinion, been disposed to lay down
"first" and most general principles of demonstration on which
all others were to hang, and then to escape any difficulties

encountered on the way by inventing such new definitions as might be required to resolve the questions at issue.

Bacon nevertheless regarded Aristotle as a man of great "wit." He first admired and later envied the Stagirite for his having at his disposal, while a relatively young man and tutor of Alexander at the Court of Macedon, a host of helpers in collecting data suitable for natural histories. Bacon also saw in Aristotle an investigator who began wisely with an observational and experimental study of nature and then foolishly forfeited the fruits of his early inquiries by turning aside to pursue "abstractions."

From Aristotle, and from Plato too, Bacon acquired the opinion that the tasks of politics are occupations appropriate to a philosopher. He welcomed Aristotle's contention that the fully virtuous citizen will have within his magnanimous disposal the products of husbandmen, mechanics, artists, and scientists. This opinion served to confirm Bacon in his congenitally expansive tastes, and to its influence may be partly ascribed his always living beyond what persons with natures less lavish than his considered more than adequate means.

Bacon left Cambridge at the close of 1575 with a reputation for extraordinary application to study. In 1576 he was enrolled at Gray's Inn. His legal studies were undertaken not for the purpose of pleading before the bar, but as a preparation for the future administration of affairs of state. These studies were interrupted less than a year after they were begun, when at the age of sixteen Bacon went to France as a member of the ambassadorial staff of Sir Amyas Paulet. There he had an opportunity to observe the complexities and intrigues of continental politics. On one occasion, he was entrusted with a diplomatic message to the Queen. An interest on his part in the causes of natural phenomena and inventiveness showed themselves at this time: he became curious about vibrations in the production of sounds, and constructed a cipher for diplomatic communication.

In 1579 Sir Nicholas Bacon was suddenly stricken and Francis was called home. The father's death was to determine the future circumstances of the son. Sir Nicholas had provided for his other children and over a period of years had been setting aside funds for the purchase of land with revenue for Francis. Death having intervened suddenly, the junior son got but one fifth of his father's personal estate.

Francis Bacon faced the world at the age of eighteen with learning unusual in one of his years, large capacity of mind, skill in the use of words, a "small portion," and great ambition. The youth reared in opulent circumstances found himself comparatively poor and dependent on wealthy relatives at Court for sustaining employment. He turned with zeal to the study of law. In 1582 he became a barrister and in 1586 a bencher. He was now bent on emulating his father by attaining high legal place and a reputation for learning and justice in dispensing the law. This ambition he was to accomplish, notwithstanding his removal from office in 1621 through political causes. Bacon's father had been well informed in the law; the son was to become the greatest authority of his time on the constitutional law of England and the possessor of a legal learning far beyond that of his contemporary Coke, who in later times was to be accounted a far greater professional lawyer and to acquire wider celebrity as a jurist.

Having prepared himself by legal studies, Bacon turned to the Court in suit for office. Burghley, to whom he early appealed, was not unmindful of the son of his brother-in-law and a former Lord Keeper. But he thought he saw in his nephew an alarming precociousness, undue self-assurance, and too great an independence of mind. There was engendered in the conservative statesman a deep-set suspicion that the nephew might not have political aptitude. This suspicion was in no degree allayed by the nephew's asking for some political office or other to "carry" him, so that through its holding he might have the "commandment" of many "wits" for the implementation of a newfangled scientific enterprise. To the early suspicion of Lord Burghley there was to be added by his heir, Lord Salisbury, jealousy toward a potential political rival. Queen Elizabeth was well disposed toward the son of a respected Lord Keeper, whom she had known from his boy-hood, even if she saw cause to banish him from her "presence" for some three years because of his promoting opposition in the Commons to her Lord Treasurer's attempt to violate the "privileges" of the Lower House by consulting the Lords on a question of supply. King James and his deputy Lord Buckingham were impressed by Bacon's capacity to sift political issues and his mastery over the House of Commons and courts of law. They accepted his services and read his "advices," if they but rarely put into effect the wise measures and shrewd procedures he advocated.

In 1584 Burghley provided his nephew with a seat in the Commons. From then on Bacon continued an influential member of that House for some thirty-six years. In 1589 he was made Clerk of the Star Chamber, but only by promise in a "reversion," and a few years thereafter Queen's Counsel without formal warrant. At the beginning of James's reign, in 1603, he was knighted, and became in succeeding years King's Counsel with patent, Solicitor-General, Clerk of the Star Chamber in fact, Judge of the Court of the Verge, Attorney-General, Lord Keeper of the Great Seal, Lord Chancellor, Baron Verulam of Verulam, and Viscount St. Albans.

Bacon sought political office both for his own support and for the furtherance of large public designs. These designs included the coherent ordering of the common and statutory laws of England, the modification of harsh and "vengeful" legal penalties, and the maintenance of the "privileges" of Parliament and courts of law against arbitrary incursions by sovereigns and their ministers. Most of all, Bacon hoped that through the influence gained by the occupancy of high office great patronage and large means would be his for the collecting of the massive natural history required to inaugurate a new inductive philosophy and to establish a new regimen of science and learning.

The holding of many public offices was in fact, however, to impede his philosophical undertaking and leave its written exposition far from complete. During a period of thirty-six years Bacon was a member of every Parliament and of nearly every important committee of the Lower House. He was the chief mediator in incessant quarrels between Commons and Lords. He held some of the most laborious political and judicial offices in government. Such writings in representation of his philosophical "instauration"—most of them incomplete, and many of them fragmentary—as he was able to set down before 1621, when he went out of political office, were prepared in brief "vacations" between sittings of Parliament and courts; those written after that year were put together in haste under apprehension that he would not have long to live.

II. *Toward a New Naturalistic Philosophy*

From the days when Bacon studied at Cambridge his mind had been occupied with a scheme of philosophy and a method of investigation which would entail a decisive break with the thinking of the past. While in early revolt against Peripatetic doctrines and practices he had turned to the dialogues of Plato and the fragments of the Pre-Platonists. In Plato's writings he discerned a "phrenetic" tendency to construct a universe out of "thoughts." But he also found examples of a rudimentary induction and a regarding of knowledge as the ascent from sensible particulars through lesser "axioms" to higher axioms, and finally to a determinate unity. He welcomed with eagerness the identification by the Pre-Platonists, especially by Democritus, of philosophy with the science of nature. In Democritus he saw a philosopher who was fortunate in being free from the doctrine of final causes and in discerning a formed and active matter, which was not the indeterminate, deprived, and inert abstraction he had met with in the writings of Plato and Aristotle.

Bacon turned also for light and guidance to the "reformers" of logic and, neglecting Platonico-Aristotelian eclectics like Pico, Pomponazzi, and Vives, looked to the professed philosophers of nature, Campanella, Cardan, Patricius, Severinus, and Telesius. He pondered the hypotheses of the new astronomers, Copernicus and Galileo, and the theories of the physicist Gilbert, and brought under review the experiments of Roger Bacon, the chemists, and the alchemists.

In Roger Bacon's experiments he saw promise of fruitful inquiry, but in the claims of the chemists and the alchemists, mainly confused empiricism employed in the service of charlatanism. Lully's method, based as it was on the mechanical alignment of some letters of the alphabet and some colors, chosen as representations of various principles, Bacon regarded as the symbolical manipulation of doctrines assumed to be already known. Lully's logical machine might be made to work but no real discovery could ever come from it. Ramus' vaunted reform of Aristotle proved on close examination to be nothing more than a specious fusion of rhetoric and logic, many of whose terms were mere metaphorical figures of speech. Astronomers were promoting a "new," heliocentric "hypothesis" which was not in fact new, for it had been proposed long before by Grecian and Roman writers, in order to "save" certain phenomena which admittedly could also be salvaged by the Ptolemaic theory of cycles and epicycles, with the earth as the center of all. These "new" astronomers could still assume, like Aristotle of antiquity, that the components of celestial bodies differed in kind from the elements composing the earth. Their failure to seek and investigate the nature, composition, and motions of one matter common both to heavenly and terrestrial bodies was, in Bacon's view, the obvious evidence of their failure as naturalists. Gilbert, Telesius, and some other recent philosophers had professed a reliance on observation and experiment and had displayed considerable capacity for inquiry, but they, like their predecessors, lacked a restraining explicit method of inductive investigation and were driven by ambition to assume that axioms which pertain to but few phenomena and hold within limited areas may be elevated, through reliance on an intemperate intellect, into a complete philosophy. These ambitious thinkers were like the overeager boy who, having come upon a tholepin on the shore, supposed he had

the makings of an entire ship! Bacon's impatience with builders of whole systems out of scant materials was characteristic
of his thinking from an early age. He gave it classic utterance
in his *Natural and Experimental History for the Foundation
of Philosophy* of 1622.

> To what purpose (he asked) are these brain-creations and
> idle displays of power? In ancient times there were philo-
> sophical doctrines in plenty; doctrines of Pythagoras, Phi-
> lolaus, Xenophanes, Heraclitus, Empedocles, Parmenides,
> Anaxagoras, Leucippus, Democritus, Plato, Aristotle, Zeno,
> and others. All these invented systems of the universe, each
> according to his own fancy, like so many arguments of
> plays; and those their inventions they recited and pub-
> lished; whereof some were more elegant and probable,
> others harsh and unlikely. Nor in our age, though by reason
> of the institutions of schools and colleges wits are more
> restrained, has the practice entirely ceased; for Patricius,
> Telesius, Brunus, Severinus the Dane, Gilbert the English-
> man, and Campanella have come upon the stage with fresh
> stories, neither honored by approbation nor elegant in argu-
> ment. Are we then to wonder at this, as if there would not
> be innumerable sects and opinions of this kind in all ages?
> There is not and never will be an end or limit to this; one
> catches at one thing, another at another; each has his favor-
> ite fancy; pure and open light there is none; everyone phi-
> losophizes out of the cells of his own imagination, as out of
> Plato's cave; the higher wits with more acuteness and felic-
> ity, the duller, less happily but with equal pertinacity. And
> now of late by the regulation of some learned and (as things
> now are) excellent men (the former variety and license
> having I suppose become wearisome), the sciences are con-
> fined to certain and prescribed authors, and thus restrained
> are imposed upon the old and instilled into the young; so
> that now (to use the sarcasm of Cicero concerning Caesar's
> year), the constellation of Lyra rises by edict, and authority
> is taken for truth, not truth for authority. Which kind of
> institution and discipline is excellent for the present use,
> but precludes all prospect of improvement. For we copy
> the sin of our first parents while we suffer for it. They
> wished to be like God, but their posterity wish to be even
> greater. For we create worlds, we direct and domineer over
> nature, we will have it that all things *are* as in our folly

we think they should be, not as seems fittest to the Divine wisdom, or as they are found to be in fact; and I know not whether we more distort the facts of nature or our own wits; but we clearly impress the stamp of our own image on the creatures and works of God, instead of carefully examining and recognizing in them the stamp of the Creator himself. Wherefore our dominion over creatures is a second time forfeited, not undeservedly; and whereas after the fall of man some power over the resistance of creatures was still left to him—the power of subduing and managing them by true and solid arts—yet this too through our insolence, and because we desire to be like God and to follow the dictates of our own reason, we in great part lose. If therefore there be any humility toward the Creator, any reverence for or disposition to magnify His works, any charity for man and anxiety to relieve his sorrows and necessities, any love of truth in nature, any hatred of darkness, any desire for the purification of the understanding, we must entreat men again and again to discard, or at least set apart for a while, these volatile and preposterous philosophies, which have preferred theses to hypotheses, led experience captive, and triumphed over the works of God; and to approach with humility and veneration to unroll the volume of Creation, to linger and meditate therein, and with minds washed clean from opinions to study it in purity and integrity. For this is that sound and language which went forth into all lands, and did not incur the confusion of Babel; this should men study to be perfect in, and becoming again as little children condescend to take the alphabet of it into their hands, and spare no pains to search and unravel the interpretation thereof, but pursue it strenuously and persevere even unto death.

It was Bacon's intention to supplant the theories of past and present schools and sects—Platonic, Peripatetic, Paracelsan, Telesian, and the rest—by a thoroughly naturalistic, materialistic philosophy, fully and not partially founded on natural history, and pursued according to the requirements of a new restraining method. The axioms or principles of this new philosophy would be statements of natural causes and natural laws derived from scientific observation and experiment, directed and interpreted according to the rules of a strict induction. This philosophy would exclude all that was transcendental and admit nothing that could be deemed a priori, unless the term transcendental could be applied to the most general of those principles confirmable, through sense observation, by data from whose examples they were derived. Its method would provide aid for the senses while controlling and purging the intellect of its overweening disposition to fly to "high priori" areas and there to remain— witness its abiding in the realm of Platonic forms dialectically sustained, and in Aristotle's Being qua Being derived through abstraction in high degree.

III. *Man and the Kingdom of Nature*

Man, the investigator of nature, is according to Bacon a natural creature with faculties by nature limited. He is also a partaker of the Divine Image. What is divine in him lies, like the will of his Maker, within the area of Divine Revelation, and beyond the purview of natural philosophy. The ethical direction of the divine part of man is to be found in the placets of revelation. As a natural creature with limited capacities befitting his nature, man cannot through his own powers attain to a knowledge of the transcendent mind and nature of God, or anything else that is divine. A metaphysics which pretends to this knowledge, like the Peripatetic ontology in which a philosophical Being qua Being and first and uncaused Cause is identified with the God of Divine Revelation or the Platonists' equation of the Divine Creator with a causal Form of the Good, is in philosophy pretension and in theology heresy. Man, according to Bacon, belongs to three kingdoms, the kingdom of God, where through divine Grace he is saved from his sins; the political kingdom in which initiative in sovereignty, justice, and law is given by God to ruling powers; and the kingdom of nature over which man at the Creation has been given dominion. For an understanding of the first and second of these kingdoms one must go to the revelation given in the Scriptures; knowledge of the third is attainable through the exercise of human faculties. Because the three kingdoms cannot be brought under one knowledge with one derivation, Bacon's philosophy is pluralistic in character. It is with the knowledge of the third kingdom, the kingdom of nature, that the *New Organon* has to do.

The subject matter of human philosophy consists, in Bacon's view, of creatures operated by natural causes, including the part of man produced in natural generation. The structures and processes of these natural creatures are physical and material. If in a new philosophy of bodies the ancient term "form" is to be retained to represent the component elements of things—and of all the terms in use it is probably the best for the purpose—the forms of this philosophy are not to be confused with Plato's occupants of a "divine" realm nor with those causes which activate and give meaning to Aristotle's potential, indeterminate matter. Plato assumed that forms, unavailable to sense, were far removed from materiate things. In his scheme they constituted, through dialectical organization, a transcendent realm of being set apart from, indeed posed in opposition to, the changing, moving world of physical particulars. Aristotle set his forms, as activating factors in natural motions, over against an indeterminate and by itself inert, meaningless, logically indiscernible matter. Bacon will make matter, whose operations are available to sense, the actual and not merely the potential stuff of nature. Bacon's forms are materiate. Matter as such is in its inherent nature formed and furnished with active and determinate characteristics, indeed with all natural causes, motions, and structure of bodies; these are none other than formed matter's manifestations.

IV. *Some Peripatetic Doctrines*

The general character, as well as the novelty, of Bacon's philosophy becomes clear when we consider his manner of rejecting the prevailing Peripatetic principles, scheme of knowledge, and methods of scientific demonstration. This rejection is everywhere manifest in his *New Organon*. The Peripatetics recognize three divisions among the sciences: the theoretical, the practical, and the productive. The end of the theoretical is contemplation; of the practical, action; of the productive, the making of things through the imposition of a secondary form on what in nature has a natural form—for example, the imposing of the form of a table on the form of a tree. This division of the sciences includes metaphysics, mathematics, and physics. These three theoretical sciences show increasing degrees of abstraction. Physics deals with the forms of materiate things in motion; mathematics with quantity in abstraction from both the matter and motion of things; and metaphysics with being in abstraction from all else. Metaphysics, called by Aristotle "first philosophy" and "theology," is according to the Peripatetics the most abstract, most inclusive, and most certain of all the sciences. Its subject matter, Being qua Being—which is also the First Cause and the Prime Mover—is convertible with a Unity founded on and established by the Principle of Identity. These three convertibles are called Transcendentals by some Peripatetics, because, as Aristotle taught, they lie beyond what is interpretable by those categories employed in demonstration within the sciences which fall below metaphysics. "Physics" in the Aristotelian, Peripatetic—and Baconian—sense extends over the whole realm of "nature" (Greek *physis*) and includes mineralogical, biological, botanical, anatomical, physiological, chemical, and psychological data.

The practical sciences are ethics and politics, in the latter of which, according to Aristotle, the former has its end. These sciences are not demonstrable, as the theoretical sciences are, because, while the subject matter of a theoretical science cannot be other than it is, contingent factors enter into the conduct of ethical and political agents, in possession of voluntary desire and choice.

In the productive division of the sciences are to be found the varied sorts of knowledge manifest in the arts—drama, medicine, agriculture, carpentry, cooking, and so on. Those arts which have to do with the body—its feeding, clothing, shelter, and the like—are by many of the Peripatetics called "mechanical," in opposition to those of a higher, intellectual sort, like music and drama.

Each science in each of the three major divisions is kept independent and separate from the others through its own distinctive underlying axioms. The axioms of one specific science are not deducible from those of another, nor from any more general axioms which might conceivably lie beyond the specific sciences. Axioms are intuitively discerned. To ask for proof of them would, according to Aristotle, be to invite a proof in turn of this proof, and then a proof of the proof of the original proof, and so on ad infinitum.

There is one science which Aristotle recognizes but omits from his classification, because he assumes that it is implied in all the sciences. This is analytic, called by some Stoics and later thinkers logic. Logic has three "first principles," the laws of identity, contradiction, and excluded middle. It includes two methods, deduction and induction. The instrument of the former is the syllogism. This consists of three propositions, so aligned that from two given propositions, called premises, a third, called the conclusion, necessarily follows. The syllogism contains three terms, two of which are joined in each proposition by a copula. In every valid syllogism the term common to the two premises, called the middle term, is employed in a universal sense at least once.

Of induction Aristotle recognizes two sorts. The first of these is "perfect" induction, which requires an exhaustive examination of all the particulars concerned. Since, except in a very limited number of rather obvious cases, this would impose an impossible operation, Aristotle does not stress it. The second kind of induction which he mentions depends on the determination by exclusion of negative instances of a universal term which represents a species. To exemplify: (C) Man, horse, mule (A) are long-lived; (C) man, horse, mule (B) are without gall; therefore, all animals without gall are long-lived. Here the conclusion obviously depends on the sup-

position that B is no wider than C—a supposition for which
Aristotle provides no proof. Bacon observes that Aristotle as
a rule arrives at his definitions of natural species through the
process of excluding contrary instances and, having done so,
makes these species fixed and "eternal," affirming the while
that whereas particulars appear and disappear the species remains and through it nature attains her ends.

Aristotle's "universe" is composed of fifty-five concentric
spheres, with the earth at the center. The outermost of these
spheres is called the *primum mobile,* the first moved by the
First Mover. The motion of the *primum mobile* is, in turn,
transmitted to other celestial bodies, whose motions may be
said to be in different degrees "imitations" of the motion of
the First Mover. Celestial motion is appropriately circular,
for the circular is the most perfect of all motions, uniform
without break, irregularity, or end. The heavenly bodies,
again, contain a quintessence or fifth element, ether, while
terrestrial bodies are composed of the four elements, fire, air,
earth, water—sometimes designated according to their fundamental qualities as the hot, the cold, the dry, and the wet.
These elements are not to be confused with Aristotle's matter,
since they are already formed "first bodies." They are not to
be thought of as existing in separation one from another.
Each of the four "first bodies," having a specific inherent
weight of its own, tends to take up its appropriate place.
Their several motions are rectilinear in character. All terrestrial motions are combinations of rectilinear motions. These
motions Aristotle classifies generally in a fourfold manner, as
local motion or change in place, increase and decrease, qualitative change, generation and decay.

Motion of whatever sort is for Aristotle a passing from
potency to act through the agency of form. Forms activate
matter, which without their agency would remain in a condition of potency, privation, indeterminateness, nonsignifi-
cance. In motion four causes are, for Aristotle, discernible:
the material, that out of which whatever in a condition of
becoming becomes; the formal, the determinate thing which

is actualized; the efficient, the operation which brings potency into act; and the final cause which is the end or purpose achieved through the passage from potency to act. Since the form is the cause which both activates matter and marks its determination and significance in actualization, Aristotle is able to reduce his four causes to two, the material and the formal. These causes produce substances. Nature consists of substances, each of which is a compound of matter and form.

V. *Bacon's Rejection of Aristotelianism*

At Bacon's hands the Peripatetic principles and arrangement of the sciences undergo a drastic reconstruction. His alternative classification of the divisions of knowledge is set forth in detail in his *Advancement of Learning* (1605) and also with some minor modifications in the Latin translation of this work, *De dignitate et augmentis scientiarum* [1623]— (*Of the Dignity and Advancement of Learning*). From his classification Bacon excludes any metaphysical ontology which has to do with a philosophical First Cause and that most abstract of all objects, a transcendental Being qua Being. The doctrine of a first, uncaused Cause Bacon considers unphilosophical for the reason that the conception first assumes the principle of cause and effect, and then—when it becomes philosophically inconvenient—deserts it. A metaphysics which takes its stand, as Aristotle's does, on a transcendental abstract unity, asserting the convertibility of this with an abstract being as such rests, in Bacon's opinion, on a circular argument which can do no more than exhibit a tautology. Aristotle himself, affirming that these transcendentals are known intuitively, significantly puts them beyond the realm of categorized things which admit of demonstration by syllogism.

Bacon identifies his metaphysics with universalized demonstrable physics. He reduces mathematics in status from an independent science to an instrument of physics. The principles of ethics, because of their having to do with the part of man which is made in the Divine Image, are placed by him under the jurisdiction of revealed theology. Knowledge in the arts, whether "intellectual" or "manual," however elevated, however lowly and commonplace, he merges with the operative part of physics. In his view all works of art, formed as they must be through the operations of matter, are as natural as stones, trees, animals, and that part of man produced in natural generation. Bacon would have it "firmly settled within the minds of men, that the artificial does not differ from the natural either in form or in essence, but only in the efficient. . . . Nor matters it, provided things are put in the way to produce an effect, whether it be done by man or apart from man."

Bacon rejects the Peripatetic principle of abstraction as a misleading guide in the organizing of objects of scientific inquiry. "One," he says, "who philosophizes rightly and orderly should dissect nature, and not abstract her." Aristotle enfranchises the science of mathematics, assigns to it abstractions separable from materiate things in motion, and argues that quantities, the objects of this science, are not existential realities. Then he goes on to stress an even more abstract science, metaphysics, whose objects lie beyond what is amenable to treatment by such categories as space, time, quantity, quality, active and passive power. But the proper objects of science are in Bacon's view concrete, material, moving things. These, he says, are the objects which admit of "dissection."

The Peripatetic separation of the sciences by means of "axioms" Bacon regards as "unscientific." It has served to perpetuate the divisions of knowledge introduced in antiquity, when the investigation of nature was scarcely begun, and has been constantly used to thwart the introduction of new investigations. It has severed the branches of knowledge, such as astronomy, optics, medicine, from the nourishing stem of general scientific principles.

Bacon regards as an unfortunate error Aristotle's belief that earthly motions and elements are different in kind from those of the celestial spheres. This opinion, which, he says, owes its origin to a pagan regard for the supposed eternal character of the heavens, has served to "corrupt" both astronomy and physics. Bacon rejects all of Aristotle's five elements.

As a philosophical materialist Bacon reduces Aristotle's four types of motion, local motion, increase and decrease, qualitative change, generation and decay, to one sort, local motion in space. Aristotle's account of motion as process through which deprived matter comes to have a form, an actualized final cause, involves in Bacon's opinion an untenable view of both matter and form. To hold, as the Peripatetics do, that potential, indeterminate matter is an ingredient in action is to affirm something which cannot be described in any positive manner. The explanation of process in terms of nonmateriate form, which is also deemed final cause, is merely a stating of what has been effected and not an account of factors operative within the action.

For Bacon forms are inherent in the matter of which moving objects are composed. A substance for him is not a conjunction of a single form and "appropriate" nondeterminate matter, but a conjunction of several materiate forms. "To inquire the form of a lion, or of an oak, of gold," says Bacon, "nay even of water or air would be to turn serious business into a game; but to inquire the form of dense, rare, hot, cold, heavy, light, tangible, pneumatic, volatile, fixed, and of similar things; and of schematisms as well as motions, which . . . are not many and yet make up and sustain the beings and forms of all substances; this, I say, it is which we are attempting, and it constitutes and defines the metaphysic of forms." Forms are the natures of their inner causes, the laws of their operations, the very things themselves. The terms form, nature, cause, and law are all convertible, each with the others. Even as lesser legal clauses are included within the more general law, so are lesser forms contained within the greater form. The lesser form stands in relation to the greater form as species stands in relation to genus. The form of heat, for instance, is motion, but only of a kind, for the motion which is heat is comprehended within a more general form of motion, far more comprehensive in operation than heat. The forms of nature, like the letters of the alphabet, are limited in number. Out of their combinations arise all things which are produced whether in natural generation or by art. In change or motion one form gives place to another form. The search for forms and demonstrations respecting their operations constitute science as theoretical; the production of works through the knowledge of them is the achievement of science as operative. Human knowledge ends in the production of works through an understanding of the materiate forms of nature.

Bacon is extremely critical of the Peripatetics' claims for their logic. They stress the unassailability of a "knowledge" derived from their "first notions," the principles of identity, contradiction, and excluded middle, and the certainty of the syllogistic demonstration which is said to follow from the acceptance of these three. Bacon, of course, as a sane man admits that what is, is—the principle of identity; that the same thing cannot both be and not be what it is—the principle of contradiction; that a thing must either be this or not this, for there is no middle alternative—the principle of excluded middle. He also acknowledges that the syllogism can play a useful role in the organizing of scientific knowledge when this has already been gained, and in the presenting of ethical, political, legal, and theological arguments—whether ill or well founded. Yet he contends vigorously that neither the "first principles" of thought nor the syllogism can furnish any new truth about nature's operations; that the syllogism demonstrates no truth not already implied in the premise which contains the universal term. Bacon condemns the promotion by the Aristotelians of the separate axioms of their several sciences to the status of primary principles. It is from these axioms as highest propositions that all lesser, "middle" scientific propositions are, in the schools, derived and attested, to the perpetuation of ancient, outmoded doctrines and divisions within knowledge and the denial of new, specific truths.

The method of scientific demonstration now taught in the

schools consists almost entirely of the syllogism. Induction is but lightly touched upon, and then forgotten. The so-called proof by syllogism, in Bacon's view, is but a relating of terms. Demonstration by syllogism is about words, not things. Its argument depends upon a universal middle term; yet for the establishment of this Aristotle and his followers provide no satisfactory direction, if indeed any direction at all. Aristotle's own middle terms are commonly vague and formed without due regard to sense and particulars. In the beginning Aristotle seems to show some respect for observation, experiment, and experience. Later on, when he comes to depend upon terms hastily defined to resolve his difficulties, he leaves experience behind, or drags her along like a captive chained to his chariot. His successors, with their elaborate definitions, have altogether deserted experience. Middle terms on which proof turns are "elected" according to every man's invention.

As means for the establishment of a middle term in a major premise Aristotle's "perfect" induction is futile. His other induction, which proceeds to definition of species by the adduction of affirmative instances to the exclusion of negative instances, does little more than exemplify what Bacon calls the "peculiar and perpetual error of the human intellect to be more moved and excited by affirmatives than by negatives." On this method Bacon makes the comment, "To conclude *upon an enumeration of particulars without instances contradictory* is no conclusion, but a conjecture; for who can assure . . . that there are not others on the contrary side which appear not." Aristotle's dependence on his second kind of induction, which determines species through selected affirmative instances only, is in Bacon's opinion the chief cause of his readily assuming that nature as such consists of a number of fixed types. Aristotle regards as unnatural things, as "monsters," those objects in "nature" which are not found to conform to his defined species. When making his physical demonstrations Aristotle concerns himself only with nature "at liberty," and excludes both nature as "vexed," when framed

into works by the arts, and nature as "impeded," when in
digression and deflection from her common motions and
generations she produces monsters. It is not surprising, then,
that those of Aristotle's successors who compile natural histories tend to segregate nature's deviations in collections of
marvels, treating them as intractable objects, strange addenda
to nature. Yet surely, says Bacon, the things which nature
herself produces are of nature, natural. The study of monsters
can surely help the scientist bent on the production of new
species and the artist eager to understand the manner in
which nature may be made to operate in the production of
new "marvels."

Bacon's own scheme of science rises, "like a triangle," from
the wide base of natural history gathered and interpreted according to the requirements of a new inductive method. Its
next stage is physics inductively established. At the top this
physics, now made general, becomes the metaphysics of
nature.

> Of Natural Philosophy (writes Bacon) the basis is Natural
> History; the stage next the basis is Physic; the stage next
> the vertical point is Metaphysic. As for the vertical point,
> *Opus quod operatur Deus a principio usque ad finem,* [the
> work which God worketh from the beginning to the end],
> the Summary Law of Nature, we know not whether man's
> inquiry can attain unto it. But these three be the true *stages*
> of knowledge; and are to them that are depraved no better
> than the giants' hills, [Pelion, Ossa, and Olympus, piled
> upon each other] . . . but to those which refer all things to
> the glory of God, they are as the three acclamations, *Sancte,*
> *sancte, sancte* [Holy, Holy, Holy]; holy in the description
> or dilatation of his works, holy in the connection or con-
> catenation of them, and holy in the union of them in a
> perpetual and uniform law. And therefore the speculation
> was excellent in Parmenides and Plato, although but a
> speculation in them. That all things by scale did ascend to
> unity. So then always that knowledge is worthiest, which
> is charged with least multiplicity; which appeareth to be
> Metaphysic; as that which considereth the Simple Forms
> or Differences of things, which are few in number, and the
> degrees and co-ordinations whereof make all this variety.

In Bacon's naturalistic scheme there is no place for a knowledge which has for its purpose mere contemplation—the activity assigned by Aristotle to his metaphysician. The aim of all knowledge is action in the production of works for the promotion of human happiness and the relief of man's estate. Through inductive science man is to recapture his dominion over nature long forfeited and long prevented through the efforts of erring philosophers and men of learning. Since knowledge is operative in design, and the acts of nature and art are one in kind, to physics there is to be assigned, as its operative part or counterpart, mechanics; and to metaphysics, which is generalized physics, magic—in the original sense of full and active wisdom. Physics will deal with things in relatively narrow contexts and implications. Metaphysics will contain what is "summary," will abridge the "circumlocutions and long courses of experience," command the "widest and most open field of operation," and having established a summary general law, form, cause, or nature will "enfranchise" man unto "the utmost possibility of superinducing that nature upon every sort of matter."

VI. *Natural History and the Instauration of the Sciences*

For the provision of help toward the collecting of natural history, the foundation on which his philosophical edifice was to rest, Bacon began suit as early as 1592 to Lord Burghley. In that year he wrote this uncle of his hope, that when a political office became his, he might "bring in industrious observations, grounded conclusions, and profitable inventions and discoveries." "I do easily see," he explained, "that place of any reasonable countenance doth bring commandment of more wits than a man's own, which is the thing I greatly affect." Three years later in a "device" or masque, presented at Gray's Inn for Elizabeth's entertainment, Bacon appealed in similar vein to the Queen herself. In 1605 he reminded James that "if Alexander made . . . liberal assignation to Aristotle of treasure for the allowance of hunters, fowlers, fishers and the like, that he might compile an History of nature, much better do they deserve it that travail in Arts of nature." In 1608 Bacon conceived the plan of acquiring an already established "place to command wits and pens, Westminster, Eton, Winchester, specially Trinity College in Cambridge, St. John's in Cambridge, Magdalene College in Oxford, and be-speaking this betimes with the King, My Lord Archbishop, My Lord Treasurer." In 1620, in his *New Organon* he implored the King's aid in collecting a natural history, and at the same time, having by now begun to despair of obtaining from members of the Court either funds or a college, he called upon his readers generally "to come forward and take part" in the work. Five years later, a year before his death, Bacon wrote to the Venetian Fulgentius, saying, "The third part of the Instauration, that is, the Natural History, it is plainly a work for a King or a Pope, or for some college or order, and it cannot be performed by private industry as it should."

During his lifetime, Bacon got no help from any public or private person for his instauration of the sciences. A year after he died his chaplain and first biographer, Rawley, wrote in a preface to what was obviously a hastily prepared collection of examples of natural history, the *Sylva sylvarum* (*Forest of Materials*):

> I have . . . heard his lordship discourse that men (no doubt) will think many of the experiments contained in this collection to be vulgar and trivial, mean and sordid, curious and fruitless. . . . I have heard his lordship speak complainingly, that his lordship (who thinketh he deserveth to be an architect in this building) should be forced to be a workman and a laborer and to dig clay and burn the brick; and more than that (according to the hard condition of the Israelites at the latter end), to gather the straw and stubble over all the fields to burn the brick withal. For he knoweth, that except he do it, nothing will be done: men are so set to despise the means of their own good.

A generation was to elapse before scientists, at home and abroad, hailing Bacon as a "new Aristotle" and "nature's sec-

retary," undertook at his bidding and according to his directions the collecting of myriad natural histories.

Bacon's Great Instauration, framed according to new principles and a new alignment of the sciences, was to contain six
parts. The first of these parts was designed for the clearing
away of the waste and rubble which, through battles among
contending philosophical sects, had accumulated about the
foundations of knowledge. The second part was to present a
new method of inquiry, hitherto unused and unknown. The
third would contain natural histories collected, arranged, and
interpreted according to the requirements of this new method.
The fourth would exhibit a "ladder" or scale of ascent in
proven knowledge, from lesser to greater axioms. The fifth
would consist of pieces of knowledge, experimentally derived
but not as yet proved and placed within a new scientific syn-
thesis. The sixth was to provide a naturalistic metaphysics or
comprehensive philosophy of nature.

The character of each of these parts, with the exception of
the first, was determined by the year 1607, when Bacon wrote
his unfinished *Outline and Argument of the Second Part of
the Instauration.* Two years earlier he had published his *Advancement of Learning.* In this he had given an account of
certain "errors," "vanities," and "oppositions" which had re-
tarded science. (Most of these are mentioned in the *New
Organon.*) He had presented his evidence in the form of a
review of the waste areas of learning, uncultivated by apt
inquiry and neglected by the professors of learning. As an
effort toward the furtherance of his scientific enterprise this
attack on past and present learning and learned institutions
had produced but little effect. The result could hardly have
been a matter for surprise to any person less sanguinely dedicated to a cause than the author. Both Elizabeth and James
had been educated in the classics. Elizabeth continued to read
Augustine. James wrote learned treatises. Burghley and Salisbury were in turn chancellors of Cambridge. The clergy and
the heads and fellows of colleges were by training and pro-
fession Platonic or Aristotelian in philosophic outlook. For-

mer attacks on traditional learning by Agrippa, Bruno, Campanella, Cardan, Telesius, and others had left learned persons and foundations unmoved.

While composing his *Advancement of Learning*, Bacon had kept a guard on his pen lest anything he wrote should prove "harsh," "out of tune," offensive to a learned sovereign and learned subjects in Court, church, and the universities. Later on when it became evident to the author that this publication was not proving an effective means for obtaining royal or other support for his instauration, Bacon pondered the wisdom of publishing a less restrained attack on traditional learning by *"discoursing scornfully* of the philosophy of the Grecians . . . *taking a greater confidence and authority* in discourses of this nature." For some years he had thought of representing the errors of past and present learning as "idols" or phantoms which perpetually beset the human mind in general and the minds of philosophers in particular. In 1620 Bacon was in his sixtieth year; nothing as yet had been accomplished either by way of exposition or exemplification of his Great Instauration. To put some of its parts, at least, "out of peril," as he said, he decided to publish his unfinished *New Organon* in representation of the second part. This would provide a new inductive method, however incomplete, a "key" to the "interpretation of nature." To the *New Organon* he would attach an introduction to the third part, a *Preparative Toward a Natural and Experimental History*. Observations and experiments recorded in the Second Book of the *New Organon* to illustrate the new method would serve also to exemplify in some degree the sort of natural history required for the third part. Since Bacon's thoughts about "idols" had at this date been set down merely in aphoristic form, and not composed as a continuous treatise, such aphorisms as were ready could be included within Book One of the *New Organon*. These aphorisms would serve to indicate what the first of the six parts of the instauration intended. As soon as the Latin version of the *Advancement of*

Learning with some deletions, modifications, and additions could be made ready, it also—since nothing better was available—would be published in representation of the first part. As soon as possible the preparation of an independent natural history would be undertaken.

By 1622 Bacon was able to produce a section of his natural history. This he published under the title *Natural and Experimental History for the Foundation of Philosophy: Or Phenomena of the Universe: Which is the Third Part of the Great Instauration.* The pressing reason for its preparation the author made clear in a "foreword." He wrote:

> It has occurred to me that there are doubtless many wits scattered over Europe, capacious, open, lofty, subtle, solid, and constant. What if one of them were to enter into the plan of my Organum and try to use it? He yet knows not what to do, nor how to prepare and address himself to the work of philosophy. If indeed it were a thing that could be accomplished by the reading of philosophical books, or discussion, or meditation, he might be equal to the work, whoever he be, and discharge it well; but if I refer him to natural history and the experiments of arts (as in fact I do), it is out of his line, he has not leisure for it, he cannot afford the expense. Yet I would not ask anyone to give up what he has until he can exchange it for something better. But when a true and copious history of nature and the arts shall have been once collected and digested, and when it shall have been set forth and unfolded before men's eyes, then will there be good hope that those great wits I spoke of before, such as flourished in the old philosophers, and are even still often to be found—wits so vigorous that out of a mere plank or shell (that is out of scanty and trifling experience) they could frame certain barks of philosophy, of admirable construction as far as the work is concerned—after they have obtained proper material and provision will raise much more solid structures; and that too though they prefer to walk on in the old path, and not by the way of my Organum, which in my estimation, if not the only, is at least the best course. It comes therefore to this, that my Organum, even if it were completed, would not without the Natural History much advance the Instauration of the sci-

ences, whereas the Natural History without the Organum would advance it not a little. And therefore, I have thought it better and wiser by all means and above all things to apply myself to this work.

Shortly after the publication of the *New Organon* Bacon was out of political office. In the remaining years of his life he did what he could to leave written works in adequate representation of the parts of his philosophic scheme. According to the "Plan of the Work" contained in the *New Organon,* the Great Instauration was to consist specifically of six parts: (1) the Division of the Sciences; (2) Directions concerning the Interpretation of Nature; (3) the Phenomena of the Universe; (4) the Ladder of the Intellect; (5) the Forerunners of the New Philosophy; and (6) the New Philosophy or Active Science. Of the fifty-odd larger works and smaller philosophical pieces composed by the author the following were in his opinion representative, in some degree, of these respective divisions: Part One: *De dignitate et augmentis scientiarum (Of the Dignity and Advancement of Learning).* Part Two: *Novum organum sive indicia de interpretatione naturae (The New Organon or True Directions Concerning the Interpretation of Nature).* Part Three: *Parasceve ad historiam naturalem et experimentalem (Preparative toward a Natural and Experimental History), Historia naturalis et experimentalis ad condendam philosophiam: sive phenomena universi (Natural and Experimental History for the Foundation of Philosophy or Phenomena of the Universe), Historia ventorum (History of the Winds), Abecedarium naturae (The Alphabet of Nature), Historia vitae et mortis (History of Life and Death), Historia densi et rari; necnon coitionis et expansionis materiae per spatia (History of Dense and Rare: the Contraction and Expansion of Matter in Space), Historia gravis et levis (History of Heavy and Light)*—a "lost work," *Historia sympathiae et antipathiae rerum (History of the Sympathy and Antipathy of Things)*—preface only, *Historia sulphuris, mercurii, et salis (History of Sulphur, Mercury, and Salt)*—

preface only, *Sylva sylvarum* (*Forest of Materials*), *Inquisitione de magnete* (*Inquiry Concerning the Loadstone*), *Topica inquisitionis de luce et lumine* (*A Topic of Inquiry Concerning Light and Illumination*). Part Four: *Scala intellectus sive filum labyrinthi* (*Ladder of the Understanding or Thread of the Labyrinth*)—preface only. Part Five: *Prodromi sive anticipationes philosophiae secundae* (*Forerunners or Anticipations of the New Philosophy*). Part Six: nil.[1]

VII. *The New Organon*

The published work of 1620, which follows the present Introduction, includes a *Proem*, an *Epistle Dedicatory*, a *Preface* to the *Great Instauration*, the *Plan of the Work*, a *Preface* to the *New Organon*, the *New Organon* proper, and a *Preparative Toward a Natural and Experimental History*. These various pieces, some brief, some lengthy, contain no statement of Bacon's classification of the sciences and only slight reference to his divorce of theological ontology from naturalistic metaphysics; yet together they may be said to contain the author's most telling statement of what is certainly a distinctive and a new philosophy. The *New Organon* represents Division Two of the Great Instauration. Of the six divisions of his instauration Bacon considered this the most important of all, and there can be no doubt that he looked upon the *New Organon* as likely to be in effect the most consequential of all the literary works produced by him in the promotion of a new learning founded on a new type of science. The method contained within it was designed to initiate the "end and termination of infinite error." Every stage of future scientific inquiry and investigation, from the collecting of natural history to the eduction of the most general metaphysical principle, was to proceed according to the requirements of this method. To the preparation of the *New*

[1] For a detailed account of these and the other philosophical works of Bacon, see the writer's *Philosophy of Francis Bacon* (Chicago, 1948), Chap. III.

Organon Bacon devoted parts of seventeen politically busy years. Rawley said that he had seen "at least twelve copies" of the work "revised year by year, one after another, and every year altered and amended in the frame thereof, till at last it came to that Model in which it was committed to the press; as many living creatures do lick their young ones till they bring them to their strength of limbs." Some of Bacon's earlier and incomplete attempts at its writing and composition remain under such titles as *Valerius Terminus of the Interpretation of Nature: with the annotations of Hermes Stella; Partis instaurationis secundae delineatio et argumentum (Outline and Argument of the Second Part of the Instauration); Filum labyrinthi, sive formula inquisitionis (Thread of the Labyrinth or Rule of Inquiry); Cogitata et visa de interpretatione naturae, sive de scientia operativa (Thoughts and Impressions: Concerning the Interpretation of Nature, or Concerning Operative Science); Aphorismi et consilia, de auxiliis mentis et accensione luminis naturalis (Aphorisms and Counsels Concerning the Mind's Aids and the Kindling of Natural Light).*

The introductory parts of the *New Organon* provide reasons for undertaking a new sort of inquiry and indicate the general character of the new induction. The First Book expounds the doctrine of Idols and advances reasons for the lack of advance in scientific knowledge. The Second Book exhibits by examples the new interpretation of nature in operation. In this the author names eleven directions for the "true interpretation of nature." These directions are designed to show "how to educe . . . axioms from experience" and in turn, "how to deduce and derive new experiments from axioms." The beginning will lie in the observations and experiments recorded in natural history. Investigation will then proceed to the discovery of the least general principles, and from this to the discovery of the more general. At each level of inquiry the axiom, whether less or more general, will, in addition to explaining the phenomena brought under observation, suggest other and more general axioms which as hypotheses—Bacon on occasion uses this term—will be tested by sense-observation in the light of particulars. The most general principles will be established last of all. They, too, like all the others, will be demonstrated by operations available to sense.

The new method of induction will help the senses, infirm and given to error as they are; it will also aid and control the intellect, ever prone to fly to first principles and there to remain, forgetful of phenomena and facts. The new organon or method will govern the knowing faculties, and will implement a "commerce between the mind and things" by promoting "forever a true and lawful marriage between the empirical and the rational faculty, the unkind and ill-starred divorce and separation of which has thrown into confusion all the affairs of the human family."

Of the eleven directions contained within Book Two of the *New Organon* Bacon calls the first "the Presentation of Instances to the Understanding" and the second "the Indulgence of the Understanding." The interpretation of nature really begins with the latter of these, which becomes the "First Vintage," so to speak, from the new vineyard of science tilled by a new kind of philosopher. Bacon illustrates his first and second directions with a search for the form or nature of heat. For the Presentation of Instances three sorts of table are to be prepared. The first of these tables will contain examples in which the nature under investigation is present, for instance, rays of the sun, flames, quicklime sprinkled with water, and substances rubbed violently together—Bacon instances some twenty-seven cases. A second table will contain instances lacking the nature under investigation, for example, the rays of the moon, and mixtures of oil and quicklime—Bacon lists some thirty-two. To these tables, one of Presence and one of Absence, a third, that of Deviation or Absence in Proximity will be added. This table will contain records of the increase and decrease of heat in the same objects. Here Bacon lists some forty-one examples, including the increase of heat in animals through exercise and in anvils through repeated blows upon their surfaces.

After making a protracted examination of the three tables of instances, Bacon offers, as the First Vintage produced through the toil of workers in the vineyard, the following definition of heat. *"Heat,"* he says, *"is a motion, expansive, restrained, and acting in its strife upon the smaller particles of bodies.* But the expansion is thus modified: *while it expands all ways,* it has at the same time an inclination *upward.* And the struggle in the particles is modified also; *it is not sluggish, but hurried and with violence."* Having in mind, now as always, his doctrine that science is in end operative, Bacon gives instructions for the production of heat. *"If,"* he says, *"in any natural body you can excite a dilating and expanding motion, and can so repress this motion and turn it back upon itself, that the dilation shall not proceed equally, but have its way in one part and be counteracted in another, you will undoubtedly generate heat;* without taking into account whether the body be elementary (as it is called) or subject to celestial influence; whether it be luminous or opaque; rare or dense; locally expanded or confined within the bounds of its first dimension; verging to dissolution or remaining in its original state; animal, vegetable, or mineral, water, oil, or air, or any other substance whatever susceptible of the above-mentioned motion."

The two directions so far illustrated are but the preliminary stages of induction. Nine other directions remain for consideration: those of Prerogative Instances—which excel "common" instances in the aiding of the senses and intellect and in the furtherance of "operation"; the Supports of Induction; the Rectification of Induction; the Variation of Inquiry according to the Nature of the Subject; Nature's Prerogative with respect to Investigation—what should be investigated first and what afterwards; the Limits of Investigation or the Synopsis of All Natures in the Universe; the Bringing Down to Practice; the Preparations for Investigation; and the Ascending and Descending Scale of Axioms.

Of Prerogative Instances, Bacon names and provides examples of twenty-seven kinds. These he separates into two groups, one having to do with the "informative," the other with the "operative" part of science. The former of these groups he subdivides into two according to their functions, the aiding of the senses and of the intellect respectively. Those Prerogative Instances which bring aid to the senses Bacon calls Instances of the Lamp—for their shedding of light. They are of five kinds: Instances of the Door or Gate, which serve to "strengthen, enlarge, and rectify" the senses; Summoning or Evoking Instances—to borrow a term from courts of law—which serve to make available to sense factors that lie concealed through circumstances; Instances of the Road or Traveling Instances, which stress especially the continuity of motion in natural objects; Supplementary Instances or Instances of Refuge which bring aid where the senses are unable to perceive objects directly; and Dissecting or Awakening Instances which are especially helpful in arousing the understanding.

As for Prerogative Instances which aid the understanding: Solitary Instances help in the exclusion of the form under investigation; Migratory, Striking, Companionship, and Subjunctive Instances indicate relatively more determinately the affirmation of the form; Clandestine, Singular, Constitutive, Conformable, Alliance, and Bordering Instances "exalt" the intellect toward the discernment of common natures and genera; Deviating Instances guide the intellect when it would be led astray through mere habit; and Instances of the Fingerpost and of Divorce warn the intellect when it might proceed in the direction of false causes.

Prerogative Instances which have to do with the operative part of science include three kinds: first, those which serve to indicate especially the aim of operation in producing "works," and the most economical means of accomplishing this—Intimating Instances and Instances of Power; secondly, those which illustrate the measure of operation—Mathematical Instances; and thirdly those which facilitate practical operation generally—Polychrest and Magical Instances.

Having illustrated Prerogative Instances in great detail and indicated which of them should be employed initially, which later on, and having then announced that he will proceed immediately to consider the remaining eight of his eleven directions, the author brings his *New Organon* to a sudden close.

Whether it was Bacon's expectation in 1620 that he would be able to continue the exposition of his inductive method at some later date is questionable. Certainly never during the remaining six years of his life, when he was working feverishly to leave works in representation of the third part of his Great Instauration, did he find time to return to this task. The *New Organon* was to remain incomplete. Thus, the reader inherits much less than half of what Bacon for years hoped and intended to bequeath to future generations.

FULTON H. ANDERSON

SELECTED BIBLIOGRAPHY

EDITIONS OF BACON'S WORKS

Works. Edited by J. Spedding, R. L. Ellis, and D. D. Heath. London, 1857. New York, 1869.

Works. Edited by J. M. Robertson. London and New York, 1905.

Novum Organum. Edited by Thomas Fowler. Oxford, 1878.

The Letters and Life of Francis Bacon. Edited and with a life by J. Spedding. London, 1861.

COLLATERAL READING

Adam, C. *Philosophie de François Bacon.* Paris, 1890.

Anderson, F. H. *The Philosophy of Francis Bacon.* Chicago, 1948.

Blunt, H. W. "Bacon's Method of Science," *Proceedings of the Aristotelian Society,* N.S. IV (1903-04), 16-31.

Boas, M. "Bacon and Gilbert," *Journal of the History of Ideas,* XII (1951), 466-7.

Bowers, R. H. "Bacon's Spider Simile," *Journal of the History of Ideas,* XVII (1956), 133-35.

Broad, C. D. *The Philosophy of Francis Bacon.* Cambridge, 1926. Also available in Broad, C. D. *Ethics and the History of Philosophy.* London, 1952.

Church, R. W. *Bacon.* ("English Men of Letters Series.") London, 1884. New York, 1894.

Dickie, W. M. "A Comparison of the Scientific Method and Achievement of Aristotle and Bacon," *Philosophical Review,* XXXI (1922), 471-94.

Dorner, A. *De Baconis Philosophia.* Berlin, 1867.

Farrington, B. *Francis Bacon: Philosopher of Industrial Science.* New York, 1949. London, 1951.

Feuerbach, Ludwig. *Geschichte der neuern Philosophie von Bacon von Verulam bis Benedict Spinoza.* Leipzig, 1847.

Fischer, K. *Francis Bacon von Verulam.* Leipzig, 1856. English translation by J. Oxenford. London, 1857.

Heussler, H. *Francis Bacon und seine geschichtliche Stellung.* Breslau, 1889.

Hochberg, H. "The Empirical Philosophy of Roger and Francis Bacon," *Philosophy of Science,* XX (1953), 313-26.

Höffding, H. *Geschichte der neuern Philosophie.* Leipzig, 1895. English translation by B. E. Meyer, London, 1900. New York, 1955.

Lasson, A. *Über Bacon von Verulam's wissenschaftliche Principien.* Berlin, 1860.

Levi, A. *Il Pensiero di Francesco Bacone.* Turin, 1925.

Levine, I. *Francis Bacon.* London, 1925.

Liebig, J. von. *Über Francis Bacon von Verulam.* Munich, 1863.

McClure, M. T. "Francis Bacon and the Modern Spirit," *Journal of Philosophy,* XIV (1917), 520-27.

McRae, R. F. "The Unity of the Sciences: Bacon, Descartes, and Leibniz," *Journal of the History of Ideas,* XVIII (1957), 27-48.

Nichol, J. *Francis Bacon, His Life and Philosophy.* Edinburgh and London, 1889.

Prior, M. E. "Bacon's Man of Science," *Journal of the History of Ideas,* XV (1954), 348-70.

de Rémusat, C. *Bacon, sa vie, son temps, sa philosophie.* Paris, 1858.

Schaub, E. L. "Francis Bacon and the Modern Spirit," *Monist,* XL (1930), 416-38.

Sorley, W. R. *A History of English Philosophy.* Cambridge, 1920.

Storck, J. "Francis Bacon and Contemporary Philosophical Difficulties," *Journal of Philosophy,* XXVIII (1931), 169-86.

Whewell, W. *On the Philosophy of Discovery.* London, 1860.

NOTE ON THE TEXT

This edition is a reprint of the standard translation of James Spedding, Robert Leslie Ellis, and Douglas Denon Heath in *The Works* (Vol. VIII), published in Boston by Taggard and Thompson in 1863. All bracketed statements are the additions of the editor.

In addition to some minor stylistic changes, spelling, punctuation, and capitalization have been revised to conform to current American usage.

THE GREAT INSTAURATION

PROEM

FRANCIS OF VERULAM

REASONED THUS WITH HIMSELF

AND JUDGED IT TO BE FOR THE INTEREST OF THE PRESENT AND FUTURE GENERATIONS THAT THEY SHOULD BE MADE ACQUAINTED WITH HIS THOUGHTS.

Being convinced that the human intellect makes its own difficulties, not using the true helps which are at man's disposal soberly and judiciously—whence follows manifold ignorance of things, and by reason of that ignorance mischiefs innumerable —he thought all trial should be made, whether that commerce between the mind of man and the nature of things, which is more precious than anything on earth, or at least than anything that is of the earth, might by any means be restored to its perfect and original condition, or if that may not be, yet reduced to a better condition than that in which it now is. Now that the errors which have hitherto prevailed, and which will prevail for ever, should (if the mind be left to go its own way) either by the natural force of the understanding or by help of the aids and instruments of logic, one by one, correct themselves, was a thing not to be hoped for, because the primary notions of things which the mind readily and passively imbibes, stores up, and accumulates (and it is from them that all the rest flow) are false, confused, and overhastily abstracted from the facts; nor are the secondary and subsequent notions less arbitrary and inconstant; whence it follows that the entire fabric of human reason which we employ in the inquisition of nature is badly put together and built up, and like some magnificent structure without any foundation. For while men are occupied in admiring and applauding the false powers of the mind, they pass by and throw away those true powers, which, if it be supplied with the proper aids and can itself be content to wait upon nature instead of vainly affecting to overrule her,

3

are within its reach. There was but one course left, therefore—to try the whole thing anew upon a better plan, and to commence a total reconstruction of sciences, arts, and all human knowledge, raised upon the proper foundations. And this, though in the project and undertaking it may seem a thing infinite and beyond all the powers of man, yet when it comes to be dealt with it will be found sound and sober, more so than what has been done hitherto. For of this there is some issue; whereas in what is now done in the matter of science there is only a whirling round about, and perpetual agitation, ending where it began. And although he was well aware how solitary an enterprise it is, and how hard a thing to win faith and credit for, nevertheless he was resolved not to abandon either it or himself, nor to be deterred from trying and entering upon that one path which is alone open to the human mind. For better it is to make a beginning of that which may lead to something, than to engage in a perpetual struggle and pursuit in courses which have no exit. And certainly the two ways of contemplation are much like those two ways of action, so much celebrated, in this—that the one, arduous and difficult in the beginning, leads out at last into the open country, while the other, seeming at first sight easy and free from obstruction, leads to pathless and precipitous places.

Moreover, because he knew not how long it might be before these things would occur to anyone else, judging especially from this, that he has found no man hitherto who has applied his mind to the like, he resolved to publish at once so much as he has been able to complete. The cause of which haste was not ambition for himself, but solicitude for the work; that in case of his death there might remain some outline and project of that which he had conceived, and some evidence likewise of his honest mind and inclination toward the benefit of the human race. Certain it is that all other ambition whatsoever seemed poor in his eyes compared with the work which he had in hand, seeing that the matter at issue is either nothing or a thing so great that it may well be content with its own merit, without seeking other recompense.

EPISTLE DEDICATORY

TO OUR MOST GRACIOUS AND MIGHTY PRINCE AND LORD

JAMES,

BY THE GRACE OF GOD
OF GREAT BRITAIN, FRANCE, AND IRELAND KING,
DEFENDER OF THE FAITH, ETC.

Most Gracious and Mighty King,

Your Majesty may perhaps accuse me of larceny, having stolen from your affairs so much time as was required for this work. I know not what to say for myself. For of time there can be no restitution unless it be that what has been abstracted from your business may perhaps go to the memory of your name and the honor of your age; if these things are indeed worth anything. Certainly they are quite new, totally new in their very kind: and yet they are copied from a very ancient model, even the world itself and the nature of things and of the mind. And to say truth, I am wont for my own part to regard this work as a child of time rather than of wit, the only wonder being that the first notion of the thing, and such great suspicions concerning matters long established, should have come into any man's mind. All the rest follows readily enough. And no doubt there is something of accident (as we call it) and luck as well in what men think as in what they do or say. But for this accident which I speak of, I wish that if there be any good in what I have to offer, it may be ascribed to the infinite mercy and goodness of God, and to the felicity of your Majesty's times; to which as I have been an honest and affectionate servant in my life, so after my death I may yet perhaps, through the kindling of this new light in the darkness of philosophy, be the means of making this age famous to posterity; and surely to the times of the wisest and most learned of kings belongs of right the regeneration and restoration of the sciences. Lastly, I have a request to make—a request no way unworthy of your Majesty, and which especially concerns the work in hand, namely, that you who resemble Solomon in so many things—in the gravity of your judgments, in the peacefulness of your reign, in the largeness of your heart, in the noble variety of the books which you have composed—would further follow his example in taking order for the collecting and perfecting of a natural and experimental history, true and severe (unincumbered with literature and book-learning), such as philosophy may be built upon—such, in fact, as I shall in its proper place describe: that so at length, after the lapse of so many ages, philosophy and the sciences may no longer float in air, but rest on the solid foundation of experience of every kind, and the same well examined and weighed. I have provided the machine, but the stuff must be gathered from the facts of nature. May God Almighty long preserve your Majesty!

> *Your Majesty's*
> *Most bounden and devoted Servant,*
> FRANCIS VERULAM,
> CHANCELLOR.

THE GREAT INSTAURATION

PREFACE

That the state of knowledge is not prosperous nor greatly advancing, and that a way must be opened for the human understanding entirely different from any hitherto known, and other helps provided, in order that the mind may exercise over the nature of things the authority which properly belongs to it.

It seems to me that men do not rightly understand either their store or their strength, but overrate the one and underrate the other. Hence it follows that either from an extravagant estimate of the value of the arts which they possess they seek no further, or else from too mean an estimate of their own powers they spend their strength in small matters and never put it fairly to the trial in those which go to the main. These are as the pillars of fate set in the path of knowledge, for men have neither desire nor hope to encourage them to penetrate further. And since opinion of store is one of the chief causes of want, and satisfaction with the present induces neglect of provision for the future, it becomes a thing not only useful, but absolutely necessary, that the excess of honor and admiration with which our existing stock of inventions is regarded be in the very entrance and threshold of the work, and that frankly and without circumlocution stripped off, and men be duly warned not to exaggerate or make too much of them. For let a man look carefully into all that variety of books with which the arts and sciences abound, he will find everywhere endless repetitions of the same thing, varying in the method of treatment, but not new in substance, insomuch that the whole stock, numerous as it appears at first view, proves on examination to be but scanty. And for its value and utility it must be plainly avowed that that wisdom which we have derived prin-

7

cipally from the Greeks is but like the boyhood of knowledge, and has the characteristic property of boys: it can talk, but it cannot generate, for it is fruitful of controversies but barren of works. So that the state of learning as it now is appears to be represented to the life in the old fable of Scylla, who had the head and face of a virgin, but her womb was hung round with barking monsters, from which she could not be delivered. For in like manner the sciences to which we are accustomed have certain general positions which are specious and flattering; but as soon as they come to particulars, which are as the parts of generation, when they should produce fruit and works, then arise contentions and barking disputations, which are the end of the matter and all the issue they can yield. Observe also, that if sciences of this kind had any life in them, that could never have come to pass which has been the case now for many ages—that they stand almost at a stay, without receiving any augmentations worthy of the human race, insomuch that many times not only what was asserted once is asserted still, but what was a question once is a question still, and instead of being resolved by discussion is only fixed and fed; and all the tradition and succession of schools is still a succession of masters and scholars, not of inventors and those who bring to further perfection the things invented. In the mechanical arts we do not find it so; they, on the contrary, as having in them some breath of life, are continually growing and becoming more perfect. As originally invented they are commonly rude, clumsy, and shapeless; afterwards they acquire new powers and more commodious arrangements and constructions, in so far that men shall sooner leave the study and pursuit of them and turn to something else than they arrive at the ultimate perfection of which they are capable. Philosophy and the intellectual sciences, on the contrary, stand like statues, worshipped and celebrated, but not moved or advanced. Nay, they sometimes flourish most in the hands of the first author, and afterwards degenerate. For when men have once made over their judgments to others' keeping, and (like those senators whom they called *Pedarii*) have agreed to support some one person's opinion, from that time they make no enlargement of the sciences themselves, but fall to the servile office of embellishing certain individual authors and increasing their retinue. And let it not be said that the sciences have been growing gradually till they have at last reached their full stature, and so (their course being completed) have settled in the works of a few writers; and that there being now no room for the invention of better, all that remains is to embellish and cultivate those things which have been invented already. Would it were so! But the truth is that this appropriating of the sciences has its origin in nothing better than the confidence of a few persons and the sloth and indolence of the rest. For after the sciences had been in several perhaps cultivated and handled diligently, there has risen up some man of bold disposition, and famous for methods and short ways which people like, who has in appearance reduced them to an art, while he has in fact only spoiled all that the others had done.[1] And yet this is what posterity likes, because it makes the work short and easy, and saves further inquiry, of which they are weary and impatient. And if any one take this general acquiescence and consent for an argument of weight, as being the judgment of Time, let me tell him that the reasoning on which he relies is most fallacious and weak. For, first, we are far from knowing all that in the matter of sciences and arts has in various ages and places been brought to light and published, much less all that has been by private persons secretly attempted and stirred; so neither the births nor the miscarriages of Time are entered in our records. Nor, secondly, is the consent itself and the time it has continued a consideration of much worth. For however various are the forms of civil polities, there is but one form of polity in the sciences; and that always has been and always will be popular. Now the doctrines which find most favor with the populace are those which are either contentious and pugnacious, or specious and empty—such, I say, as either entangle assent or tickle it. And therefore no doubt the greatest wits in each successive age have been forced out of their own course:

1 [Reference is to Aristotle. Cf. the editor's Introduction.]

men of capacity and intellect above the vulgar having been fain, for reputation's sake, to bow to the judgment of the time and the multitude; and thus if any contemplations of a higher order took light anywhere, they were presently blown out by the winds of vulgar opinions. So that Time is like a river which has brought down to us things light and puffed up, while those which are weighty and solid have sunk. Nay, those very authors who have usurped a kind of dictatorship in the sciences and taken upon them to lay down the law with such confidence, yet when from time to time they come to themselves again, they fall to complaints of the subtlety of nature, the hiding places of truth, the obscurity of things, the entanglement of causes, the weakness of the human mind; wherein nevertheless they show themselves never the more modest, seeing that they will rather lay the blame upon the common condition of men and nature than upon themselves. And then whatever any art fails to attain, they ever set it down upon the authority of that art itself as impossible of attainment; and how can art be found guilty when it is judge in its own cause? So it is but a device for exempting ignorance from ignominy. Now for those things which are delivered and received, this is their condition: barren of works, full of questions; in point of enlargement slow and languid, carrying a show of perfection in the whole, but in the parts ill filled up; in selection popular, and unsatisfactory even to those who propound them; and therefore fenced round and set forth with sundry artifices. And if there be any who have determined to make trial for themselves and put their own strength to the work of advancing the boundaries of the sciences, yet have they not ventured to cast themselves completely loose from received opinions or to seek their knowledge at the fountain; but they think they have done some great thing if they do but add and introduce into the existing sum of science something of their own, prudently considering with themselves that by making the addition they can assert their liberty, while they retain the credit of modesty by assenting to the rest. But these mediocrities and middle ways so much praised, in deferring to opinions and customs, turn to the great detriment of the sciences. For it is hardly possible at once to admire an author and to go beyond him, knowledge being as water, which will not rise above the level from which it fell. Men of this kind, therefore, amend some things, but advance little, and improve the condition of knowledge, but do not extend its range. Some, indeed, there have been who have gone more boldly to work and, taking it all for an open matter and giving their genius full play, have made a passage for themselves and their own opinions by pulling down and demolishing former ones; and yet all their stir has but little advanced the matter, since their aim has been not to extend philosophy and the arts in substance and value, but only to change doctrines and transfer the kingdom of opinions to themselves; whereby little has indeed been gained, for though the error be the opposite of the other, the causes of erring are the same in both. And if there have been any who, not binding themselves either to other men's opinions or to their own, but loving liberty, have desired to engage others along with themselves in search, these, though honest in intention, have been weak in endeavor. For they have been content to follow probable reasons and are carried round in a whirl of arguments, and in the promiscuous liberty of search have relaxed the severity of inquiry. There is none who has dwelt upon experience and the facts of nature as long as is necessary. Some there are indeed who have committed themselves to the waves of experience and almost turned mechanics, yet these again have in their very experiments pursued a kind of wandering inquiry, without any regular system of operations. And besides they have mostly proposed to themselves certain petty tasks, taking it for a great matter to work out some single discovery—a course of proceeding at once poor in aim and unskillful in design. For no man can rightly and successfully investigate the nature of anything in the thing itself; let him vary his experiments as laboriously as he will, he never comes to a resting-place, but still finds something to seek beyond. And there is another thing to be remembered— namely, that all industry in experimenting has begun with pro-

posing to itself certain definite works to be accomplished, and
has pursued them with premature and unseasonable eagerness;
it has sought, I say, experiments of fruit, not experiments of
light, not imitating the divine procedure, which in its first
day's work created light only and assigned to it one entire day,
on which day it produced no material work, but proceeded to
that on the days following. As for those who have given the
first place to logic, supposing that the surest helps to the sciences were to be found in that, they have indeed most truly
and excellently perceived that the human intellect left to its
own course is not to be trusted; but then the remedy is altogether too weak for the disease, nor is it without evil in itself.
For the logic which is received, though it be very properly
applied to civil business and to those arts which rest in discourse and opinion, is not nearly subtle enough to deal with
nature; and in attempting what it cannot master, has done
more to establish and perpetuate error than to open the way
to truth.

Upon the whole, therefore, it seems that men have not been
happy hitherto either in the trust which they have placed in
others or in their own industry with regard to the sciences;
especially as neither the demonstrations nor the experiments
as yet known are much to be relied upon. But the universe to
the eye of the human understanding is framed like a labyrinth,
presenting as it does on every side so many ambiguities of way,
such deceitful resemblances of objects and signs, natures so
irregular in their lines and so knotted and entangled. And
then the way is still to be made by the uncertain light of the
sense, sometimes shining out, sometimes clouded over, through
the woods of experience and particulars; while those who offer
themselves for guides are (as was said) themselves also puzzled,
and increase the number of errors and wanderers. In circumstances so difficult neither the natural force of man's judgment
nor even any accidental felicity offers any chance of success.
No excellence of wit, no repetition of chance experiments, can
overcome such difficulties as these. Our steps must be guided
by a clue, and the whole way from the very first perception of

the senses must be laid out upon a sure plan. Not that I would be understood to mean that nothing whatever has been done in so many ages by so great labors. We have no reason to be ashamed of the discoveries which have been made, and no doubt the ancients proved themselves in everything that turns on wit and abstract meditation, wonderful men. But, as in former ages, when men sailed only by observation of the stars, they could indeed coast along the shores of the old continent or cross a few small and Mediterranean seas; but before the ocean could be traversed and the new world discovered, the use of the mariner's needle, as a more faithful and certain guide, had to be found out; in like manner the discoveries which have been hitherto made in the arts and sciences are such as might be made by practice, meditation, observation, argumentation—for they lay near to the senses and immediately beneath common notions; but before we can reach the remoter and more hidden parts of nature, it is necessary that a more perfect use and application of the human mind and intellect be introduced.

For my own part at least, in obedience to the everlasting love of truth, I have committed myself to the uncertainties and difficulties and solitudes of the ways and, relying on the divine assistance, have upheld my mind both against the shocks and embattled ranks of opinion, and against my own private and inward hesitations and scruples, and against the fogs and clouds of nature, and the phantoms flitting about on every side, in the hope of providing at last for the present and future generations guidance more faithful and secure. Wherein if I have made any progress, the way has been opened to me by no other means than the true and legitimate humiliation of the human spirit. For all those who before me have applied themselves to the invention of arts have but cast a glance or two upon facts and examples and experience, and straightway proceeded, as if invention were nothing more than an exercise of thought, to invoke their own spirits to give them oracles. I, on the contrary, dwelling purely and constantly among the facts of nature, withdraw my intellect from them no further

than may suffice to let the images and rays of natural objects
meet in a point, as they do in the sense of vision; whence it
follows that the strength and excellence of the wit has but little to do in the matter. And the same humility which I use in
inventing I employ likewise in teaching. For I do not endeavor
either by triumphs of confutation, or pleadings of antiquity,
or assumption of authority, or even by the veil of obscurity,
to invest these inventions of mine with any majesty; which
might easily be done by one who sought to give luster to his
own name rather than light to other men's minds. I have not
sought (I say) nor do I seek either to force or ensnare men's
judgments, but I lead them to things themselves and the con-
cordances of things, that they may see for themselves what they
have, what they can dispute, what they can add and contribute
to the common stock. And for myself, if in anything I have
been either too credulous or too little awake and attentive, or
if I have fallen off by the way and left the inquiry incomplete,
nevertheless I so present these things naked and open, that
my errors can be marked and set aside before the mass of
knowledge be further infected by them; and it will be easy
also for others to continue and carry on my labors. And by
these means I suppose that I have established forever a true
and lawful marriage between the empirical and the rational
faculty, the unkind and ill-starred divorce and separation of
which has thrown into confusion all the affairs of the human
family.

Wherefore, seeing that these things do not depend upon
myself, at the outset of the work I most humbly and fervently
pray to God the Father, God the Son, and God the Holy
Ghost, that remembering the sorrows of mankind and the
pilgrimage of this our life wherein we wear out days few and
evil, they will vouchsafe through my hands to endow the
human family with new mercies. This likewise I humbly pray,
that things human may not interfere with things divine, and
that from the opening of the ways of sense and the increase of
natural light there may arise in our minds no incredulity or

darkness with regard to the divine mysteries, but rather that the understanding being thereby purified and purged of fancies and vanity, and yet not the less subject and entirely submissive to the divine oracles, may give to faith that which is faith's. Lastly, that knowledge being now discharged of that venom which the serpent infused into it, and which makes the mind of man to swell, we may not be wise above measure and sobriety, but cultivate truth in charity.

And now, having said my prayers, I turn to men, to whom I have certain salutary admonitions to offer and certain fair requests to make. My first admonition (which was also my prayer) is that men confine the sense within the limits of duty in respect of things divine: for the sense is like the sun, which reveals the face of earth, but seals and shuts up the face of heaven. My next, that in flying from this evil they fall not into the opposite error, which they will surely do if they think that the inquisition of nature is in any part interdicted or forbidden. For it was not that pure and uncorrupted natural knowledge whereby Adam gave names to the creatures according to their propriety, which gave occasion to the fall. It was the ambitious and proud desire of moral knowledge to judge of good and evil, to the end that man may revolt from God and give laws to himself, which was the form and manner of the temptation. Whereas of the sciences which regard nature, the divine philosopher declares that "it is the glory of God to conceal a thing, but it is the glory of the King to find a thing out." Even as though the divine nature took pleasure in the innocent and kindly sport of children playing at hide-and-seek, and vouchsafed of his kindness and goodness to admit the human spirit for his playfellow at that game. Lastly, I would address one general admonition to all—that they consider what are the true ends of knowledge, and that they seek it not either for pleasure of the mind, or for contention, or for superiority to others, or for profit, or fame, or power, or any of these inferior things, but for the benefit and use of life, and that they perfect and govern it in charity. For it was from lust of power that the angels fell, from lust of knowledge that man fell; but of charity there can be no excess, neither did angel or man ever come in danger by it.

The requests I have to make are these. Of myself I say nothing; but in behalf of the business which is in hand I entreat men to believe that it is not an opinion to be held, but a work to be done; and to be well assured that I am laboring to lay the foundation, not of any sect or doctrine, but of human utility and power. Next, I ask them to deal fairly by their own interests, and laying aside all emulations and prejudices in favor of this or that opinion, to join in consultation for the common good; and being now freed and guarded by the securities and helps which I offer from the errors and impediments of the way, to come forward themselves and take part in that which remains to be done. Moreover, to be of good hope, nor to imagine that this Instauration of mine is a thing infinite and beyond the power of man, when it is in fact the true end and termination of infinite error; and seeing also that it is by no means forgetful of the conditions of mortality and humanity (for it does not suppose that the work can be altogether completed within one generation, but provides for its being taken up by another); and finally that it seeks for the sciences not arrogantly in the little cells of human wit, but with reverence in the greater world. But it is the empty things that are vast; things solid are most contracted and lie in little room. And now I have only one favor more to ask (else injustice to me may perhaps imperil the business itself)—that men will consider well how far, upon that which I must needs assert (if I am to be consistent with myself), they are entitled to judge and decide upon these doctrines of mine; inasmuch as all that premature human reasoning which anticipates inquiry, and is abstracted from the facts rashly and sooner than is fit, is by me rejected (so far as the inquisition of nature is concerned) as a thing uncertain, confused, and ill built up; and I cannot be fairly asked to abide by the decision of a tribunal which is itself on trial.

THE PLAN OF THE GREAT INSTAURATION

The Instauration includes six Parts:

1. The Divisions of the Sciences
2. The New Organon; or Directions concerning the Interpretation of Nature
3. The Phenomena of the Universe; or a Natural and Experimental History for the Foundation of Philosophy
4. The Ladder of the Intellect
5. The Forerunners; or Anticipations of the New Philosophy
6. The New Philosophy; or Active Science

The Arguments of the Several Parts

It being part of my design to set everything forth, as far as may be, plainly and perspicuously (for nakedness of the mind is still, as nakedness of the body once was, the companion of innocence and simplicity), let me first explain the order and plan of the work. I distribute it into six parts.

The first part exhibits a summary or general description of the knowledge which the human race at present possesses. For I thought it good to make some pause upon that which is received; that thereby the old may be more easily made perfect and the new more easily approached. And I hold the improvement of that which we have to be as much an object as the acquisition of more. Besides which it will make me the better listened to; for " He that is ignorant (says the proverb) receives not the words of knowledge, unless thou first tell him that which is in his own heart." We will therefore make a coasting voyage along the shores of the arts and sciences received, not without importing into them some useful things by the way.

In laying out the divisions of the sciences, however, I take into account not only things already invented and known, but likewise things omitted which ought to be there. For there are found in the intellectual as in the terrestrial globe waste regions as well as cultivated ones. It is no wonder, therefore, if I am sometimes obliged to depart from the ordinary divisions. For in adding to the total you necessarily alter the parts and sections; and the received divisions of the sciences are fitted only to the received sum of them as it stands now.

With regard to those things which I shall mark as omitted, I intend not merely to set down a simple title or a concise argument of that which is wanted. For as often as I have occasion to report anything as deficient, the nature of which is at all obscure, so that men may not perhaps easily understand what I mean or what the work is which I have in my head, I shall always (provided it be a matter of any worth) take care to subjoin either directions for the execution of such work, or else a portion of the work itself executed by myself as a sample of the whole, thus giving assistance in every case either by work or by counsel. For if it were for the sake of my own reputation only and other men's interests were not concerned in it, I would not have any man think that in such cases merely some light and vague notion has crossed my mind, and that the things which I desire and attempt are no better than wishes, when they are in fact things which men may certainly command if they will, and of which I have formed in my own mind a clear and detailed conception. For I do not propose merely to survey these regions in my mind, like an augur taking auspices, but to enter them like a general who means to take possession. So much for the first part of the work.

Having thus coasted past the ancient arts, the next point is to equip the intellect for passing beyond. To the second part, therefore, belongs the doctrine concerning the better and more perfect use of human reason in the inquisition of things, and the true helps of the understanding, that thereby (as far as the condition of mortality and humanity allows) the intellect may be raised and exalted, and made capable of overcoming the difficulties and obscurities of nature. The art which I introduce with this view (which I call "Interpretation of Nature ") is a kind of logic, though the difference between it and the ordinary logic is great, indeed, immense. For the ordinary logic professes to contrive and prepare helps and guards for the understanding, as mine does; and in this one point they agree. But mine differs from it in three points especially—viz., in the end aimed at, in the order of demonstration, and in the starting point of the inquiry.

For the end which this science of mine proposes is the invention not of arguments but of arts; not of things in accordance with principles, but of principles themselves; not of probable reasons, but of designations and directions for works. And as the intention is different, so, accordingly, is the effect; the effect of the one being to overcome an opponent in argument, of the other to command nature in action.

In accordance with this end is also the nature and order of the demonstrations. For in the ordinary logic almost all the work is spent about the syllogism. Of induction, the logicians seem hardly to have taken any serious thought, but they pass it by with a slight notice and hasten on to the formulæ of disputation. I, on the contrary, reject demonstration by syllogism as acting too confusedly and letting nature slip out of its hands. For although no one can doubt that things which agree in a middle term agree with one another (which is a proposition of mathematical certainty), yet it leaves an opening for deception, which is this: the syllogism consists of propositions—propositions of words; and words are the tokens and signs of notions. Now if the very notions of the mind (which are as the soul of words and the basis of the whole structure) be improperly and overhastily abstracted from facts, vague, not sufficiently definite, faulty—in short, in many ways, the whole edifice tumbles. I therefore reject the syllogism, and that not only as regards principles (for to principles the logicians themselves do not apply it) but also as regards middle propositions, which, though obtainable no doubt by the syllogism, are, when so obtained, barren of works, remote from practice, and altogether unavailable for the active department of the sciences. Although, therefore, I leave to the syllogism and these famous and boasted modes of demonstration their jurisdiction over popular arts and such as are matter of opinion (in which department I leave all as it is), yet in dealing with the nature of things I use induction throughout, and that in the minor propositions as well as the major. For I consider induction to be that form of demonstration which upholds the sense, and closes with nature, and comes to the very brink of operation, if it does not actually deal with it.

Hence it follows that the order of demonstration is likewise inverted. For hitherto the proceeding has been to fly at once from the sense and particulars up to the most general propositions, as certain fixed poles for the argument to turn upon, and from these to derive the rest by middle terms—a short way, no doubt, but precipitate and one which will never lead to nature, though it offers an easy and ready way to disputation. Now my plan is to proceed regularly and gradually from one axiom to another, so that the most general are not reached till the last; but then, when you do come to them, you find them to be not empty notions but well defined, and such as nature would really recognize as her first principles, and such as lie at the heart and marrow of things.

But the greatest change I introduce is in the form itself of induction and the judgment made thereby. For the induction of which the logicians speak, which proceeds by simple enumeration, is a puerile thing, concludes at hazard, is always liable to be upset by a contradictory instance, takes into account only what is known and ordinary, and leads to no result.

Now what the sciences stand in need of is a form of induction which shall analyze experience and take it to pieces, and by a due process of exclusion and rejection lead to an inevitable conclusion. And if that ordinary mode of judgment practiced by the logicians was so laborious, and found exercise for such great wits, how much more labor must we be prepared to bestow upon this other, which is extracted not merely out of the depths of the mind, but out of the very bowels of nature.

Nor is this all. For I also sink the foundations of the sciences deeper and firmer; and I begin the inquiry nearer the source than men have done heretofore, submitting to examination those things which the common logic takes on trust. For first, the logicians borrow the principles of each science from the science itself; secondly, they hold in reverence the first notions of the mind; and lastly, they receive as conclusive the immediate informations of the sense, when well disposed. Now upon the first point, I hold that true logic ought to enter the several provinces of science armed with a higher authority than belongs to the principles of those sciences themselves, and ought to call those putative principles to account until they are fully established. Then with regard to the first notions of the intellect, there is not one of the impressions taken by the intellect when left to go its own way, but I hold it as suspect and no way established until it has submitted to a new trial and a fresh judgment has been thereupon pronounced. And lastly, the information of the sense itself I sift and examine in many ways. For certain it is that the senses deceive; but then at the same time they supply the means of discovering their own errors; only the errors are here, the means of discovery are to seek.

The sense fails in two ways. Sometimes it gives no information, sometimes it gives false information. For first, there are very many things which escape the sense, even when best disposed and no way obstructed, by reason either of the subtlety of the whole body or the minuteness of the parts, or distance of place, or slowness or else swiftness of motion, or familiarity of the object, or other causes. And again when the sense does apprehend a thing its apprehension is not much to be relied upon. For the testimony and information of the sense has reference always to man, not to the universe; and it is a great error to assert that the sense is the measure of things.

To meet these difficulties, I have sought on all sides diligently and faithfully to provide helps for the sense—substitutes to supply its failures, rectifications to correct its errors; and this I endeavor to accomplish not so much by instruments as by experiments. For the subtlety of experiments is far greater than that of the sense itself, even when assisted by exquisite instruments—such experiments, I mean, as are skillfully and artificially devised for the express purpose of determining the point in question. To the immediate and proper perception of the sense, therefore, I do not give much weight; but I contrive that the office of the sense shall be only to judge of the experiment, and that the experiment itself shall judge of the thing. And thus I conceive that I perform the office of a true priest of the sense (from which all knowledge in nature must be sought, unless men mean to go mad) and a not unskillful interpreter of its oracles; and that while others only profess to uphold and cultivate the sense, I do so in fact. Such then are the provisions I make for finding the genuine light of nature and kindling and bringing it to bear. And they would be sufficient of themselves if the human intellect were even and like a fair sheet of paper with no writing on it. But since the minds of men are strangely possessed and beset so that there is no true and even surface left to reflect the genuine rays of things, it is necessary to seek a remedy for this also.

Now the idols, or phantoms, by which the mind is occupied are either adventitious or innate. The adventitious come into the mind from without—namely, either from the doctrines and sects of philosophers or from perverse rules of demonstration. But the innate are inherent in the very nature of the intellect, which is far more prone to error than the sense is. For let men please themselves as they will in admiring and almost adoring the human mind, this is certain: that as an uneven mirror distorts the rays of objects according to its own figure and section, so the mind, when it receives impressions of objects through the sense, cannot be trusted to report them truly, but in forming its notions mixes up its own nature with the nature of things.

And as the first two kinds of idols are hard to eradicate, so idols of this last kind cannot be eradicated at all. All that can be done is to point them out, so that this insidious action of the mind may be marked and reproved (else as fast as old errors are destroyed new ones will spring up out of the ill complexion of the mind itself, and so we shall have but a change of errors, and not a clearance); and to lay it down once for all as a fixed and established maxim that the intellect is not qualified to judge except by means of induction, and induction in its legitimate form. This doctrine, then, of the expurgation of the intellect to qualify it for dealing with truth is comprised in three refutations: the refutation of the philosophies; the refutation of the demonstrations; and the refutation of the natural human reason. The explanation of which things, and of the true relation between the nature of things and the nature of the mind, is as the strewing and decoration of the bridal chamber of the mind and the universe, the divine goodness assisting, out of which marriage let us hope (and be this the prayer of the bridal song) there may spring helps to man, and a line and race of inventions that may in some degree subdue and overcome the necessities and miseries of humanity. This is the second part of the work.

But I design not only to indicate and mark out the ways, but also to enter them. And therefore the third part of the work embraces the "phenomena of the universe"; that is to say, experience of every kind, and such a natural history as may serve for a foundation to build philosophy upon. For a good method of demonstration or form of interpreting nature may keep the mind from going astray or stumbling, but it is not any excellence of method that can supply it with the material of knowledge. Those, however, who aspire not to guess and divine, but to discover and know, who propose not to devise mimic and fabulous worlds of their own, but to examine and dissect the nature of this very world itself, must go to facts themselves for everything. Nor can the place of this labor and search and world-wide perambulation be supplied by any genius or meditation or argumentation; no, not if all men's wits could meet in one. This, therefore, we must have or the business must be forever abandoned. But up to this day such has been the condition of men in this matter that it is no wonder if nature will not give herself into their hands.

For first, the information of the sense itself, sometimes failing, sometimes false; observation, careless, irregular, and led by chance; tradition, vain, and fed on rumor; practice, slavishly bent upon its work; experiment, blind, stupid, vague, and prematurely broken off; lastly, natural history trivial and poor—all these have contributed to supply the understanding with very bad materials for philosophy and the sciences.

Then an attempt is made to mend the matter by a preposterous subtlety and winnowing of argument. But this comes too late, the case being already past remedy, and is far from setting the business right or sifting away the errors. The only hope, therefore, of any greater increase or progress lies in a reconstruction of the sciences.

Of this reconstruction the foundation must be laid in natural history, and that of a new kind and gathered on a new principle. For it is in vain that you polish the mirror if there are no images to be reflected; and it is as necessary that the intellect should be supplied with fit matter to work upon, as with safeguards to guide its working. But my history differs from that in use (as my logic does) in many things—in end and office, in mass and composition, in subtlety, in selection also, and setting forth, with a view to the operations which are to follow.

For first, the object of the natural history which I propose is not so much to delight with variety of matter or to help with present use of experiments, as to give light to the discovery of causes and supply a suckling philosophy with its first food. For though it be true that I am principally in pursuit of works and the active department of the sciences, yet I wait for harvest-time and do not attempt to mow the moss or to reap the green corn. For I well know that axioms once rightly discovered will carry whole troops of works along with them, and produce them, not here and there one, but in clusters. And that unseasonable and puerile hurry to snatch by way of earnest at the first works which come within reach, I utterly condemn and reject as an Atalanta's apple that hinders the race.[1] Such then is the office of this natural history of mine.

Next, with regard to the mass and composition of it: I mean it to be a history not only of nature free and at large (when she is left to her own course and does her work her own way)—such as that of the heavenly bodies, meteors, earth and sea, minerals, plants, animals—but much more of nature under constraint and vexed; that is to say, when by art and the hand of man she is forced out of her natural state, and squeezed and moulded. Therefore I set down at length all experiments of the mechanical arts, of the operative part of the liberal arts, of the many crafts which have not yet grown into arts properly so called, so far as I have been able to examine them and as they conduce to the end in view. Nay (to say the plain truth), I do in fact (low and vulgar as men may think it) count more upon this part both for helps and safeguards than upon the other, seeing that the nature of things betrays itself more readily under the vexations of art than in its natural freedom.

Nor do I confine the history to bodies, but I have thought it my duty besides to make a separate history of such virtues as may be considered cardinal in nature. I mean those original passions or desires of matter which constitute the primary elements of nature; such as dense and rare, hot and cold, solid and fluid, heavy and light, and several others.

Then again, to speak of subtlety: I seek out and get together a kind of experiments much subtler and simpler than those which occur accidentally. For I drag into light many things which no one who was not proceeding by a regular and certain way to the discovery of causes would have thought of inquiring

[1] [Reference is here to Atalanta of Greek legend, who challenged her suitors to a race. She would marry only the man who could defeat her. Hippomenes (or Melanion) accepted the challenge and, on the advice of Aphrodite, dropped three golden apples on the way. Atalanta could not resist picking them up and thus lost the race.—Ed.]

after, being indeed in themselves of no great use; which shows
that they were not sought for on their own account, but having
just the same relation to things and works which the letters of
the alphabet have to speech and words—which, though in
themselves useless, are the elements of which all discourse is
made up.

Further, in the selection of the relation and experiments I
conceive I have been a more cautious purveyor than those who
have hitherto dealt with natural history. For I admit nothing
but on the faith of eyes, or at least of careful and severe examination, so that nothing is exaggerated for wonder's sake,
but what I state is sound and without mixture of fables or
vanity. All received or current falsehoods also (which by
strange negligence have been allowed for many ages to prevail
and become established) I proscribe and brand by name, that
the sciences may be no more troubled with them. For it has
been well observed that the fables and superstitions and follies
which nurses instill into children do serious injury to their
minds; and the same consideration makes me anxious, having
the management of the childhood, as it were, of philosophy
in its course of natural history, not to let it accustom itself in
the beginning to any vanity. Moreover, whenever I come to a
new experiment of any subtlety (though it be in my own
opinion certain and approved), I nevertheless subjoin a clear
account of the manner in which I made it, that men, knowing
exactly how each point was made out, may see whether there
be any error connected with it and may arouse themselves to
devise proofs more trustworthy and exquisite, if such can be
found; and finally, I interpose everywhere admonitions and
scruples and cautions, with a religious care to eject, repress,
and, as it were, exorcise every kind of phantasm.

Lastly, knowing how much the sight of man's mind is distracted by experience and history, and how hard it is at the
first (especially for minds either tender or preoccupied) to become familiar with nature, I not unfrequently subjoin observations of my own, being as the first offers inclinations, and, as
it were, glances of history toward philosophy, both by way of

an assurance to men that they will not be kept forever tossing on the waves of experience, and also that when the time comes for the intellect to begin its work, it may find everything the more ready. By such a natural history, then, as I have described, I conceive that a safe and convenient approach may be made to nature, and matter supplied of good quality and well prepared for the understanding to work upon.

And now that we have surrounded the intellect with faithful helps and guards, and got together with most careful selection a regular army of divine works, it may seem that we have no more to do but to proceed to philosophy itself. And yet in a matter so difficult and doubtful there are still some things which it seems necessary to premise, partly for convenience of explanation, partly for present use.

Of these the first is to set forth examples of inquiry and invention according to my method, exhibited by anticipation in some particular subjects; choosing such subjects as are at once the most noble in themselves among those under inquiry, and most different one from another, that there may be an example in every kind. I do not speak of those examples which are joined to the several precepts and rules by way of illustration (for of these I have given plenty in the second part of the work); but I mean actual types and models, by which the entire process of the mind and the whole fabric and order of invention from the beginning to the end, in certain subjects, and those various and remarkable, should be set, as it were, before the eyes. For I remember that in the mathematics it is easy to follow the demonstration when you have a machine beside you, whereas without that help all appears involved and more subtle than it really is. To examples of this kind— being in fact nothing more than an application of the second part in detail and at large—the fourth part of the work is devoted.

The fifth part is for temporary use only, pending the completion of the rest, like interest payable from time to time until the principal be forthcoming. For I do not make so blindly for the end of my journey as to neglect anything useful that may turn up by the way. And therefore I include in this part such things as I have myself discovered, proved, or added —not, however, according to the true rules and methods of interpretation, but by the ordinary use of the understanding in inquiring and discovering. For besides that I hope my speculations may, in virtue of my continual conversancy with nature, have a value beyond the pretensions of my wit, they will serve in the meantime for wayside inns, in which the mind may rest and refresh itself on its journey to more certain conclusions. Nevertheless I wish it to be understood in the meantime that they are conclusions by which (as not being discovered and proved by the true form of interpretation) I do not at all mean to bind myself. Nor need any one be alarmed at such suspension of judgment in one who maintains not simply that nothing can be known, but only that nothing can be known except in a certain course and way; and yet establishes provisionally certain degrees of assurance for use and relief until the mind shall arrive at a knowledge of causes in which it can rest. For even those schools of philosophy which held the absolute impossibility of knowing anything were not inferior to those which took upon them to pronounce. But then they did not provide helps for the sense and understanding, as I have done, but simply took away all their authority; which is quite a different thing—almost the reverse.

The sixth part of my work (to which the rest is subservient and ministrant) discloses and sets forth that philosophy which by the legitimate, chaste, and severe course of inquiry which I have explained and provided is at length developed and established. The completion, however, of this last part is a thing both above my strength and beyond my hopes. I have made a beginning of the work—a beginning, as I hope, not unimportant: the fortune of the human race will give the issue, such an issue, it may be, as in the present condition of things and men's minds cannot easily be conceived or imagined. For the matter in hand is no mere felicity of speculation, but the real business and fortunes of the human race, and all power of operation. For man is but the servant and interpreter of nature: what he does and what he knows is only what he has observed of nature's order in fact or in thought; beyond this he knows nothing and can do nothing. For the chain of causes cannot by any force be loosed or broken, nor can nature be commanded except by being obeyed. And so those twin objects, human knowledge and human power, do really meet in one; and it is from ignorance of causes that operation fails.

And all depends on keeping the eye steadily fixed upon the facts of nature and so receiving their images simply as they are. For God forbid that we should give out a dream of our own imagination for a pattern of the world; rather may he graciously grant to us to write an apocalypse or true vision of the footsteps of the Creator imprinted on his creatures.

Therefore do thou, O Father, who gavest the visible light as the first fruits of creation, and didst breathe into the face of man the intellectual light as the crown and consummation thereof, guard and protect this work, which coming from thy goodness returneth to thy glory. Thou when thou turnedst to look upon the works which thy hands had made, sawest that all was very good, and didst rest from thy labors. But man, when he turned to look upon the work which his hands had made, saw that all was vanity and vexation of spirit, and could find no rest therein. Wherefore if we labor in thy works with the sweat of our brows, thou wilt make us partakers of thy vision and thy sabbath. Humbly we pray that this mind may be steadfast in us, and that through these our hands, and the hands of others to whom thou shalt give the same spirit, thou wilt vouchsafe to endow the human family with new mercies.

THE NEW ORGANON

OR TRUE DIRECTIONS CONCERNING
THE INTERPRETATION OF NATURE

AUTHOR'S PREFACE

Those who have taken upon them to lay down the law of nature as a thing already searched out and understood, whether they have spoken in simple assurance or professional affectation, have therein done philosophy and the sciences great injury. For as they have been successful in inducing belief, so they have been effective in quenching and stopping inquiry; and have done more harm by spoiling and putting an end to other men's efforts than good by their own. Those on the other hand who have taken a contrary course, and asserted that absolutely nothing can be known—whether it were from hatred of the ancient sophists, or from uncertainty and fluctuation of mind, or even from a kind of fullness of learning, that they fell upon this opinion—have certainly advanced reasons for it that are not to be despised; but yet they have neither started from true principles nor rested in the just conclusion, zeal and affectation having carried them much too far. The more ancient of the Greeks (whose writings are lost) took up with better judgment a position between these two extremes—between the presumption of pronouncing on everything, and the despair of comprehending anything; and though frequently and bitterly complaining of the difficulty of inquiry and the obscurity of things, and like impatient horses champing at the bit, they did not the less follow up their object and engage with nature, thinking (it seems) that this very question—viz., whether or not anything can be known—was to be settled not by arguing, but by trying. And yet they too, trusting entirely to the force of their understanding, applied no rule, but made everything turn upon hard thinking and perpetual working and exercise of the mind.

Now my method, though hard to practice, is easy to explain; and it is this. I propose to establish progressive stages of certainty. The evidence of the sense, helped and guarded by a

certain process of correction, I retain. But the mental operation
which follows the act of sense I for the most part reject; and
instead of it I open and lay out a new and certain path for the
mind to proceed in, starting directly from the simple sensu-
ous perception. The necessity of this was felt, no doubt, by
those who attributed so much importance to logic, showing
thereby that they were in search of helps for the understanding, and had no confidence in the native and spontaneous
process of the mind. But this remedy comes too late to do any
good, when the mind is already, through the daily intercourse
and conversation of life, occupied with unsound doctrines and
beset on all sides by vain imaginations. And therefore that art
of logic, coming (as I said) too late to the rescue, and no way
able to set matters right again, has had the effect of fixing errors rather than disclosing truth. There remains but one
course for the recovery of a sound and healthy condition—
namely, that the entire work of the understanding be com-
menced afresh, and the mind itself be from the very outset not
left to take its own course, but guided at every step; and the
business be done as if by machinery. Certainly if in things
mechanical men had set to work with their naked hands, without help or force of instruments, just as in things intellectual
they have set to work with little else than the naked forces of
the understanding, very small would the matters have been
which, even with their best efforts applied in conjunction,
they could have attempted or accomplished. Now (to pause a
while upon this example and look in it as in a glass) let us
suppose that some vast obelisk were (for the decoration of a
triumph or some such magnificence) to be removed from its
place, and that men should set to work upon it with their
naked hands, would not any sober spectator think them mad?
And if they should then send for more people, thinking that
in that way they might manage it, would he not think them
all the madder? And if they then proceeded to make a selection, putting away the weaker hands, and using only the
strong and vigorous, would he not think them madder than
ever? And if lastly, not content with this, they resolved to call

in aid the art of athletics, and required all their men to come with hands, arms, and sinews well anointed and medicated according to the rules of the art, would he not cry out that they were only taking pains to show a kind of method and discretion in their madness? Yet just so it is that men proceed in matters intellectual—with just the same kind of mad effort and useless combination of forces—when they hope great things either from the number and cooperation or from the excellency and acuteness of individual wits; yea, and when they endeavor by logic (which may be considered as a kind of athletic art) to strengthen the sinews of the understanding, and yet with all this study and endeavor it is apparent to any true judgment that they are but applying the naked intellect all the time; whereas in every great work to be done by the hand of man it is manifestly impossible, without instruments and machinery, either for the strength of each to be exerted or the strength of all to be united.

Upon these premises two things occur to me of which, that they may not be overlooked, I would have men reminded. First, it falls out fortunately as I think for the allaying of contradictions and heartburnings, that the honor and reverence due to the ancients remains untouched and undiminished, while I may carry out my designs and at the same time reap the fruit of my modesty. For if I should profess that I, going the same road as the ancients, have something better to produce, there must needs have been some comparison or rivalry between us (not to be avoided by any art of words) in respect of excellency or ability of wit; and though in this there would be nothing unlawful or new (for if there be anything misapprehended by them, or falsely laid down, why may not I, using a liberty common to all, take exception to it?) yet the contest, however just and allowable, would have been an unequal one perhaps, in respect of the measure of my own powers. As it is, however (my object being to open a new way for the understanding, a way by them untried and unknown), the case is altered: party zeal and emulation are at an end, and I appear merely as a guide to point out the road—an of-

fice of small authority, and depending more upon a kind of
luck than upon any ability or excellency. And thus much relates to the persons only. The other point of which I would
have men reminded relates to the matter itself.

Be it remembered then that I am far from wishing to interfere with the philosophy which now flourishes, or with any
other philosophy more correct and complete than this which
has been or may hereafter be propounded. For I do not object to the use of this received philosophy, or others like it,
for supplying matter for disputations or ornaments for discourse—for the professor's lecture and for the business of life.
Nay, more, I declare openly that for these uses the philosophy
which I bring forward will not be much available. It does not
lie in the way. It cannot be caught up in passage. It does not
flatter the understanding by conformity with preconceived
notions. Nor will it come down to the apprehension of the
vulgar except by its utility and effects.

Let there be therefore (and may it be for the benefit of
both) two streams and two dispensations of knowledge, and in
like manner two tribes or kindreds of students in philosophy—
tribes not hostile or alien to each other, but bound together
by mutual services; let there in short be one method for the
cultivation, another for the invention, of knowledge.

And for those who prefer the former, either from hurry or
from considerations of business or for want of mental power
to take in and embrace the other (which must needs be most
men's case), I wish that they may succeed to their desire in
what they are about, and obtain what they are pursuing. But
if there be any man who, not content to rest in and use the
knowledge which has already been discovered, aspires to penetrate further; to overcome, not an adversary in argument, but
nature in action; to seek, not pretty and probable conjectures,
but certain and demonstrable knowledge—I invite all such to
join themselves, as true sons of knowledge, with me, that
passing by the outer courts of nature, which numbers have
trodden, we may find a way at length into her inner chambers. And to make my meaning clearer and to familiarize the

thing by giving it a name, I have chosen to call one of these methods or ways *Anticipation of the Mind,* the other *Interpretation of Nature.*

Moreover, I have one request to make. I have on my own part made it my care and study that the things which I shall propound should not only be true, but should also be presented to men's minds, how strangely soever preoccupied and obstructed, in a manner not harsh or unpleasant. It is but reasonable, however (especially in so great a restoration of learning and knowledge), that I should claim of men one favor in return, which is this: if anyone would form an opinion or judgment either out of his own observation, or out of the crowd of authorities, or out of the forms of demonstration (which have now acquired a sanction like that of judicial laws), concerning these speculations of mine, let him not hope that he can do it in passage or by the by; but let him examine the thing thoroughly; let him make some little trial for himself of the way which I describe and lay out; let him familiarize his thoughts with that subtlety of nature to which experience bears witness; let him correct by seasonable patience and due delay the depraved and deep-rooted habits of his mind; and when all this is done and he has begun to be his own master, let him (if he will) use his own judgment.

APHORISMS

[BOOK ONE]

I

Man, being the servant and interpreter of Nature, can do and understand so much and so much only as he has observed in fact or in thought of the course of nature. Beyond this he neither knows anything nor can do anything.

II

Neither the naked hand nor the understanding left to itself can effect much. It is by instruments and helps that the work is done, which are as much wanted for the understanding as for the hand. And as the instruments of the hand either give motion or guide it, so the instruments of the mind supply either suggestions for the understanding or cautions.

III

Human knowledge and human power meet in one; for where the cause is not known the effect cannot be produced. Nature to be commanded must be obeyed; and that which in contemplation is as the cause is in operation as the rule.

IV

Toward the effecting of works, all that man can do is to put together or put asunder natural bodies. The rest is done by nature working within.

V

The study of nature with a view to works is engaged in by the mechanic, the mathematician, the physician, the alchemist, and the magician; but by all (as things now are) with slight endeavor and scanty success.

VI

It would be an unsound fancy and self-contradictory to expect that things which have never yet been done can be done except by means which have never yet been tried.

VII

The productions of the mind and hand seem very numerous in books and manufactures. But all this variety lies in an exquisite subtlety and derivations from a few things already known, not in the number of axioms.

VIII

Moreover, the works already known are due to chance and experiment rather than to sciences; for the sciences we now possess are merely systems for the nice ordering and setting forth of things already invented, not methods of invention or directions for new works.

IX

The cause and root of nearly all evils in the sciences is this —that while we falsely admire and extol the powers of the human mind we neglect to seek for its true helps.

X

The subtlety of nature is greater many times over than the subtlety of the senses and understanding; so that all those specious meditations, speculations, and glosses in which men indulge are quite from the purpose, only there is no one by to observe it.

XI

As the sciences which we now have do not help us in finding out new works, so neither does the logic which we now have help us in finding out new sciences.

XII

The logic now in use serves rather to fix and give stability to the errors which have their foundation in commonly received notions than to help the search after truth. So it does more harm than good.

XIII

The syllogism is not applied to the first principles of sciences, and is applied in vain to intermediate axioms, being no match for the subtlety of nature. It commands assent therefore to the proposition, but does not take hold of the thing.

XIV

The syllogism consists of propositions, propositions consist of words, words are symbols of notions. Therefore if the notions themselves (which is the root of the matter) are confused and overhastily abstracted from the facts, there can be no firmness in the superstructure. Our only hope therefore lies in a true induction.

XV

There is no soundness in our notions, whether logical or physical. Substance, Quality, Action, Passion, Essence itself, are not sound notions; much less are Heavy, Light, Dense, Rare, Moist, Dry, Generation, Corruption, Attraction, Repulsion, Element, Matter, Form, and the like; but all are fantastical and ill defined.

XVI

Our notions of less general species, as Man, Dog, Dove, and of the immediate perceptions of the sense, as Hot, Cold, Black, White, do not materially mislead us; yet even these are sometimes confused by the flux and alteration of matter and the mixing of one thing with another. All the others which men have hitherto adopted are but wanderings, not being abstracted and formed from things by proper methods.

XVII

Nor is there less of willfulness and wandering in the construction of axioms than in the formation of notions, not excepting even those very principles which are obtained by common induction; but much more in the axioms and lower propositions educed by the syllogism.

XVIII

The discoveries which have hitherto been made in the sciences are such as lie close to vulgar notions, scarcely beneath the surface. In order to penetrate into the inner and further recesses of nature, it is necessary that both notions and axioms be derived from things by a more sure and guarded way, and that a method of intellectual operation be introduced altogether better and more certain.

XIX

There are and can be only two ways of searching into and discovering truth. The one flies from the senses and particulars to the most general axioms, and from these principles, the truth of which it takes for settled and immovable, proceeds to judgment and to the discovery of middle axioms. And this way is now in fashion. The other derives axioms from the senses and particulars, rising by a gradual and unbroken ascent, so that it arrives at the most general axioms last of all. This is the true way, but as yet untried.

XX

The understanding left to itself takes the same course (namely, the former) which it takes in accordance with logical order. For the mind longs to spring up to positions of higher generality, that it may find rest there, and so after a little while wearies of experiment. But this evil is increased by logic, because of the order and solemnity of its disputations.

XXI

The understanding left to itself, in a sober, patient, and grave mind, especially if it be not hindered by received doctrines, tries a little that other way, which is the right one, but with little progress, since the understanding, unless directed and assisted, is a thing unequal, and quite unfit to contend with the obscurity of things.

XXII

Both ways set out from the senses and particulars, and rest in the highest generalities; but the difference between them is infinite. For the one just glances at experiment and particulars in passing, the other dwells duly and orderly among them.

The one, again, begins at once by establishing certain abstract and useless generalities, the other rises by gradual steps to that which is prior and better known in the order of nature.

XXIII

There is a great difference between the Idols of the human mind and the Ideas of the divine. That is to say, between certain empty dogmas, and the true signatures and marks set upon the works of creation as they are found in nature.

XXIV

It cannot be that axioms established by argumentation should avail for the discovery of new works, since the subtlety of nature is greater many times over than the subtlety of argument. But axioms duly and orderly formed from particulars easily discover the way to new particulars, and thus render sciences active.

XXV

The axioms now in use, having been suggested by a scanty and manipular experience and a few particulars of most general occurrence, are made for the most part just large enough to fit and take these in; and therefore it is no wonder if they do not lead to new particulars. And if some opposite instance, not observed or not known before, chance to come in the way, the axiom is rescued and preserved by some frivolous distinction; whereas the truer course would be to correct the axiom itself.

XXVI

The conclusions of human reason as ordinarily applied in matters of nature, I call for the sake of distinction *Anticipations of Nature* (as a thing rash or premature). That reason which is elicited from facts by a just and methodical process, I call *Interpretation of Nature.*

XXVII

Anticipations are a ground sufficiently firm for consent, for even if men went mad all after the same fashion, they might agree one with another well enough.

XXVIII

For the winning of assent, indeed, anticipations are far more powerful than interpretations, because being collected from a few instances, and those for the most part of familiar occurrence, they straightway touch the understanding and fill the imagination; whereas interpretations, on the other hand, being gathered here and there from very various and widely dispersed facts, cannot suddenly strike the understanding; and therefore they must needs, in respect of the opinions of the time, seem harsh and out of tune, much as the mysteries of faith do.

XXIX

In sciences founded on opinions and dogmas, the use of anticipations and logic is good; for in them the object is to command assent to the proposition, not to master the thing.

XXX

Though all the wits of all the ages should meet together and combine and transmit their labors, yet will no great progress ever be made in science by means of anticipations; because radical errors in the first concoction of the mind are not to be cured by the excellence of functions and subsequent remedies.

XXXI

It is idle to expect any great advancement in science from the superinducing and engrafting of new things upon old. We must begin anew from the very foundations, unless we would revolve forever in a circle with mean and contemptible progress.

XXXII

The honor of the ancient authors, and indeed of all, remains untouched, since the comparison I challenge is not of wits or faculties, but of ways and methods, and the part I take upon myself is not that of a judge, but of a guide.

XXXIII

This must be plainly avowed: no judgment can be rightly formed either of my method or of the discoveries to which it leads, by means of anticipations (that is to say, of the reasoning which is now in use); since I cannot be called on to abide by the sentence of a tribunal which is itself on trial.

XXXIV

Even to deliver and explain what I bring forward is no easy matter, for things in themselves new will yet be apprehended with reference to what is old.

XXXV

It was said by Borgia of the expedition of the French into Italy, that they came with chalk in their hands to mark out their lodgings, not with arms to force their way in. I in like manner would have my doctrine enter quietly into the minds that are fit and capable of receiving it; for confutations cannot be employed when the difference is upon first principles and very notions, and even upon forms of demonstration.

XXXVI

One method of delivery alone remains to us which is simply this: we must lead men to the particulars themselves, and their series and order; while men on their side must force themselves for a while to lay their notions by and begin to familiarize themselves with facts.

XXXVII

The doctrine of those who have denied that certainty could be attained at all has some agreement with my way of proceeding at the first setting out; but they end in being infinitely separated and opposed. For the holders of that doctrine assert simply that nothing can be known. I also assert that not much can be known in nature by the way which is now in use. But then they go on to destroy the authority of the senses and understanding; whereas I proceed to devise and supply helps for the same.

XXXVIII

The idols and false notions which are now in possession of the human understanding, and have taken deep root therein, not only so beset men's minds that truth can hardly find entrance, but even after entrance is obtained, they will again in the very instauration of the sciences meet and trouble us, unless men being forewarned of the danger fortify themselves as far as may be against their assaults.

XXXIX

There are four classes of Idols which beset men's minds. To these for distinction's sake I have assigned names, calling the first class *Idols of the Tribe;* the second, *Idols of the Cave;* the third, *Idols of the Market Place;* the fourth, *Idols of the Theater.*

XL

The formation of ideas and axioms by true induction is no doubt the proper remedy to be applied for the keeping off and clearing away of idols. To point them out, however, is of great use; for the doctrine of Idols is to the interpretation of nature what the doctrine of the refutation of sophisms is to common logic.

XLI

The Idols of the Tribe have their foundation in human nature itself, and in the tribe or race of men. For it is a false assertion that the sense of man is the measure of things. On the contrary, all perceptions as well of the sense as of the mind are according to the measure of the individual and not according to the measure of the universe. And the human understanding is like a false mirror, which, receiving rays irregularly, distorts and discolors the nature of things by mingling its own nature with it.

XLII

The Idols of the Cave are the idols of the individual man. For everyone (besides the errors common to human nature in general) has a cave or den of his own, which refracts and discolors the light of nature, owing either to his own proper and peculiar nature; or to his education and conversation with others; or to the reading of books, and the authority of those whom he esteems and admires; or to the differences of impressions, accordingly as they take place in a mind preoccupied and predisposed or in a mind indifferent and settled; or the like. So that the spirit of man (according as it is meted out to different individuals) is in fact a thing variable and full of perturbation, and governed as it were by chance. Whence it was well observed by Heraclitus that men look for sciences in their own lesser worlds, and not in the greater or common world.

XLIII

There are also Idols formed by the intercourse and association of men with each other, which I call Idols of the Market Place, on account of the commerce and consort of men there. For it is by discourse that men associate, and words are imposed according to the apprehension of the vulgar. And therefore the ill and unfit choice of words wonderfully obstructs the understanding. Nor do the definitions or explanations wherewith in some things learned men are wont to guard and defend themselves, by any means set the matter right. But words plainly force and overrule the understanding, and throw all into confusion, and lead men away into numberless empty controversies and idle fancies.

XLIV

Lastly, there are Idols which have immigrated into men's minds from the various dogmas of philosophies, and also from wrong laws of demonstration. These I call Idols of the Theater, because in my judgment all the received systems are but so many stage plays, representing worlds of their own creation after an unreal and scenic fashion. Nor is it only of the systems now in vogue, or only of the ancient sects and philosophies, that I speak; for many more plays of the same kind may yet be composed and in like artificial manner set forth; seeing that errors the most widely different have nevertheless causes for the most part alike. Neither again do I mean this only of entire systems, but also of many principles and axioms in science, which by tradition, credulity, and negligence have come to be received.

But of these several kinds of Idols I must speak more largely and exactly, that the understanding may be duly cautioned.

XLV

The human understanding is of its own nature prone to suppose the existence of more order and regularity in the world than it finds. And though there be many things in nature which are singular and unmatched, yet it devises for them parallels and conjugates and relatives which do not exist. Hence the fiction that all celestial bodies move in perfect circles, spirals and dragons being (except in name) utterly rejected. Hence too the element of fire with its orb is brought in, to make up the square with the other three which the sense perceives. Hence also the ratio of density of the so-called elements is arbitrarily fixed at ten to one. And so on of other dreams. And these fancies affect not dogmas only, but simple notions also.

XLVI

The human understanding when it has once adopted an opinion (either as being the received opinion or as being agreeable to itself) draws all things else to support and agree with it. And though there be a greater number and weight of instances to be found on the other side, yet these it either neglects and despises, or else by some distinction sets aside and rejects, in order that by this great and pernicious predetermination the authority of its former conclusions may remain inviolate. And therefore it was a good answer that was made by one who, when they showed him hanging in a temple a picture of those who had paid their vows as having escaped shipwreck, and would have him say whether he did not now acknowledge the power of the gods—"Aye," asked he again, "but where are they painted that were drowned after their vows?" And such is the way of all superstition, whether in astrology, dreams, omens, divine judgments, or the like;

wherein men, having a delight in such vanities, mark the events where they are fulfilled, but where they fail, though this happen much oftener, neglect and pass them by. But with far more subtlety does this mischief insinuate itself into philosophy and the sciences; in which the first conclusion colors and brings into conformity with itself all that come after, though far sounder and better. Besides, independently of that delight and vanity which I have described, it is the peculiar and perpetual error of the human intellect to be more moved and excited by affirmatives than by negatives; whereas it ought properly to hold itself indifferently disposed toward both alike. Indeed, in the establishment of any true axiom, the negative instance is the more forcible of the two.

XLVII

The human understanding is moved by those things most which strike and enter the mind simultaneously and suddenly, and so fill the imagination; and then it feigns and supposes all other things to be somehow, though it cannot see how, similar to those few things by which it is surrounded. But for that going to and fro to remote and heterogeneous instances by which axioms are tried as in the fire, the intellect is altogether slow and unfit, unless it be forced thereto by severe laws and overruling authority.

XLVIII

The human understanding is unquiet; it cannot stop or rest, and still presses onward, but in vain. Therefore it is that we cannot conceive of any end or limit to the world, but always as of necessity it occurs to us that there is something beyond. Neither, again, can it be conceived how eternity has flowed down to the present day, for that distinction which is commonly received of infinity in time past and in time to come can by no means hold; for it would thence follow that one infinity is greater than another, and that infinity is wasting away and tending to become finite. The like subtlety arises touching the infinite divisibility of lines, from the same inability of thought to stop. But this inability interferes more mischievously in the discovery of causes; for although the most general principles in nature ought to be held merely positive, as they are discovered, and cannot with truth be referred to a cause, nevertheless the human understanding being unable to rest still seeks something prior in the order of nature. And then it is that in struggling toward that which is further off it falls back upon that which is nearer at hand, namely, on final causes, which have relation clearly to the nature of man rather than to the nature of the universe; and from this source have strangely defiled philosophy. But he is no less an unskilled and shallow philosopher who seeks causes of that which is most general, than he who in things subordinate and subaltern omits to do so.

XLIX

The human understanding is no dry light, but receives an infusion from the will and affections; whence proceed sciences which may be called "sciences as one would." For what a man had rather were true he more readily believes. Therefore he rejects difficult things from impatience of research; sober things, because they narrow hope; the deeper things of nature, from superstition; the light of experience, from arrogance and pride, lest his mind should seem to be occupied with things mean and transitory; things not commonly believed, out of deference to the opinion of the vulgar. Numberless, in short, are the ways, and sometimes imperceptible, in which the affections color and infect the understanding.

L

But by far the greatest hindrance and aberration of the human understanding proceeds from the dullness, incompetency, and deceptions of the senses; in that things which strike the sense outweigh things which do not immediately strike it, though they be more important. Hence it is that speculation commonly ceases where sight ceases; insomuch that of things invisible there is little or no observation. Hence all the working of the spirits enclosed in tangible bodies lies hid and unobserved of men. So also all the more subtle changes of form in the parts of coarser substances (which they commonly call alteration, though it is in truth local motion through exceedingly small spaces) is in like manner unobserved. And yet unless these two things just mentioned be searched out and brought to light, nothing great can be achieved in nature, as far as the production of works is concerned. So again the essential nature of our common air, and of all bodies less dense than air (which are very many), is almost unknown. For the sense by itself is a thing infirm and erring; neither can instruments for enlarging or sharpening the senses do much; but all the truer kind of interpretation of nature is effected by instances and experiments fit and apposite; wherein the sense decides touching the experiment only, and the experiment touching the point in nature and the thing itself.

LI

The human understanding is of its own nature prone to abstractions and gives a substance and reality to things which are fleeting. But to resolve nature into abstractions is less to our purpose than to dissect her into parts; as did the school of Democritus, which went further into nature than the rest. Matter rather than forms should be the object of our attention, its configurations and changes of configuration, and simple action, and law of action or motion; for forms are figments of the human mind, unless you will call those laws of action forms.

LII

Such then are the idols which I call *Idols of the Tribe,* and which take their rise either from the homogeneity of the substance of the human spirit, or from its preoccupation, or from its narrowness, or from its restless motion, or from an infusion of the affections, or from the incompetency of the senses, or from the mode of impression.

LIII

The *Idols of the Cave* take their rise in the peculiar constitution, mental or bodily, of each individual; and also in education, habit, and accident. Of this kind there is a great number and variety. But I will instance those the pointing out of which contains the most important caution, and which have most effect in disturbing the clearness of the understanding.

LIV

Men become attached to certain particular sciences and speculations, either because they fancy themselves the authors and inventors thereof, or because they have bestowed the greatest pains upon them and become most habituated to them. But men of this kind, if they betake themselves to philosophy and contemplation of a general character, distort and color them in obedience to their former fancies; a thing especially to be noticed in Aristotle, who made his natural philosophy a mere bond servant to his logic, thereby rendering it contentious and well-nigh useless. The race of chemists, again out of a few experiments of the furnace, have built up a fantastic philosophy, framed with reference to a few things; and Gilbert also, after he had employed himself most laboriously in the study and observation of the loadstone, proceeded at once to construct an entire system in accordance with his favorite subject.

LV

There is one principal and as it were radical distinction between different minds, in respect of philosophy and the sciences, which is this: that some minds are stronger and apter to mark the differences of things, others to mark their resemblances. The steady and acute mind can fix its contemplations and dwell and fasten on the subtlest distinctions; the lofty and discursive mind recognizes and puts together the finest and most general resemblances. Both kinds, however, easily err in excess, by catching the one at gradations, the other at shadows.

LVI

There are found some minds given to an extreme admiration of antiquity, others to an extreme love and appetite for novelty; but few so duly tempered that they can hold the mean, neither carping at what has been well laid down by the ancients, nor despising what is well introduced by the moderns. This, however, turns to the great injury of the sciences and philosophy, since these affectations of antiquity and novelty are the humors of partisans rather than judgments; and truth is to be sought for not in the felicity of any age, which is an unstable thing, but in the light of nature and experience, which is eternal. These factions therefore must be abjured, and care must be taken that the intellect be not hurried by them into assent.

LVII

Contemplations of nature and of bodies in their simple form break up and distract the understanding, while contemplations of nature and bodies in their composition and configuration overpower and dissolve the understanding, a distinction well seen in the school of Leucippus and Democritus as compared with the other philosophies. For that school is so busied with the particles that it hardly attends to the structure, while the others are so lost in admiration of the structure that they do not penetrate to the simplicity of nature. These kinds of contemplation should therefore be alternated and taken by turns, so that the understanding may be rendered at once penetrating and comprehensive, and the inconveniences above mentioned, with the idols which proceed from them, may be avoided.

LVIII

Let such then be our provision and contemplative prudence for keeping off and dislodging the *Idols of the Cave*, which grow for the most part either out of the predominance of a favorite subject, or out of an excessive tendency to compare or to distinguish, or out of partiality for particular ages, or out of the largeness or minuteness of the objects contemplated. And generally let every student of nature take this as a rule: that whatever his mind seizes and dwells upon with peculiar satisfaction is to be held in suspicion, and that so much the more care is to be taken in dealing with such questions to keep the understanding even and clear.

LIX

But the *Idols of the Market Place* are the most troublesome of all—idols which have crept into the understanding through the alliances of words and names. For men believe that their reason governs words; but it is also true that words react on the understanding; and this it is that has rendered philosophy and the sciences sophistical and inactive. Now words, being commonly framed and applied according to the capacity of the vulgar, follow those lines of division which are most obvious to the vulgar understanding. And whenever an understanding of greater acuteness or a more diligent observation would alter those lines to suit the true divisions of nature, words stand in the way and resist the change. Whence it comes to pass that the high and formal discussions of learned men end oftentimes in disputes about words and names; with which (according to the use and wisdom of the mathematicians) it would be more prudent to begin, and so by means of definitions reduce them to order. Yet even definitions cannot cure this evil in dealing with natural and material things, since the definitions themselves consist of words, and those words beget others. So that it is necessary to recur to individual instances, and those in due series and order, as I shall say presently when I come to the method and scheme for the formation of notions and axioms.

LX

The idols imposed by words on the understanding are of two kinds. They are either names of things which do not exist (for as there are things left unnamed through lack of observation, so likewise are there names which result from fantastic suppositions and to which nothing in reality corresponds), or they are names of things which exist, but yet confused and ill-defined, and hastily and irregularly derived from realities. Of the former kind are Fortune, the Prime Mover, Planetary Orbits, Element of Fire, and like fictions which owe their origin to false and idle theories. And this class of idols is more easily expelled, because to get rid of them it is only necessary that all theories should be steadily rejected and dismissed as obsolete.

But the other class, which springs out of a faulty and unskillful abstraction, is intricate and deeply rooted. Let us take for example such a word as *humid* and see how far the several things which the word is used to signify agree with each other, and we shall find the word *humid* to be nothing else than a mark loosely and confusedly applied to denote a variety of actions which will not bear to be reduced to any constant meaning. For it both signifies that which easily spreads itself round any other body; and that which in itself is indeterminate and cannot solidize; and that which readily yields in every direction; and that which easily divides and scatters itself; and that which easily unites and collects itself; and that which readily flows and is put in motion; and that which readily clings to another body and wets it; and that which is easily reduced to a liquid, or being solid easily melts. Accordingly, when you come to apply the word, if you take it in one sense, flame is humid; if in another, air is not humid; if in another, fine dust is humid; if in another, glass is humid. So that it is easy to see that the notion is taken by abstraction only from water and common and ordinary liquids, without any due verification.

There are, however, in words certain degrees of distortion and error. One of the least faulty kinds is that of names of substances, especially of lowest species and well-deduced (for the notion of *chalk* and of *mud* is good, of *earth* bad); a more faulty kind is that of actions, as *to generate, to corrupt, to alter;* the most faulty is of qualities (except such as are the immediate objects of the sense) as *heavy, light, rare, dense,* and the like. Yet in all these cases some notions are of necessity a little better than others, in proportion to the greater variety of subjects that fall within the range of the human sense.

LXI

But the *Idols of the Theater* are not innate, nor do they steal into the understanding secretly, but are plainly impressed and received into the mind from the playbooks of philosophical systems and the perverted rules of demonstration. To attempt refutations in this case would be merely inconsistent with what I have already said, for since we agree neither upon principles ncr upon demonstrations there is no place for argument. And this is so far well, inasmuch as it leaves the honor of the ancients untouched. For they are no wise disparaged—the question between them and me being only as to the way. For as the saying is, the lame man who keeps the right road outstrips the runner who takes a wrong one. Nay, it is obvious that when a man runs the wrong way, the more active and swift he is, the further he will go astray.

But the course I propose for the discovery of sciences is such as leaves but little to the acuteness and strength of wits, but places all wits and understandings nearly on a level. For as in the drawing of a straight line or a perfect circle, much depends on the steadiness and practice of the hand, if it be done by aim of hand only, but if with the aid of rule or compass, little or nothing; so is it exactly with my plan. But though particular confutations would be of no avail, yet touching the sects and general divisions of such systems I must say something; something also touching the external signs which show that they are unsound; and finally something touching the causes of such great infelicity and of such lasting and general agreement in error; that so the access to truth may be made less difficult, and the human understanding may the more willingly submit to its purgation and dismiss its idols.

LXII

Idols of the Theater, or of Systems, are many, and there can be and perhaps will be yet many more. For were it not that now for many ages men's minds have been busied with religion and theology; and were it not that civil governments, especially monarchies, have been averse to such novelties, even in matters speculative; so that men labor therein to the peril and harming of their fortunes—not only unrewarded, but exposed also to contempt and envy—doubtless there would have arisen many other philosophical sects like those which in great variety flourished once among the Greeks. For as on the phenomena of the heavens many hypotheses may be constructed, so likewise (and more also) many various dogmas may be set up and established on the phenomena of philosophy. And in the plays of this philosophical theater you may observe the same thing which is found in the theater of the poets, that stories invented for the stage are more compact and elegant, and more as one would wish them to be, than true stories out of history.

In general, however, there is taken for the material of philosophy either a great deal out of a few things, or a very little out of many things; so that on both sides philosophy is based on too narrow a foundation of experiment and natural history, and decides on the authority of too few cases. For the

Rational School of philosophers snatches from experience a variety of common instances, neither duly ascertained nor diligently examined and weighed, and leaves all the rest to meditation and agitation of wit.

There is also another class of philosophers who, having bestowed much diligent and careful labor on a few experiments, have thence made bold to educe and construct systems, wresting all other facts in a strange fashion to conformity therewith.

And there is yet a third class, consisting of those who out of faith and veneration mix their philosophy with theology and traditions; among whom the vanity of some has gone so far aside as to seek the origin of sciences among spirits and genii. So that this parent stock of errors—this false philosophy—is of three kinds: the Sophistical, the Empirical, and the Superstitious.

LXIII

The most conspicuous example of the first class was Aristotle, who corrupted natural philosophy by his logic: fashioning the world out of categories; assigning to the human soul, the noblest of substances, a genus from words of the second intention; doing the business of density and rarity (which is to make bodies of greater or less dimensions, that is, occupy greater or less spaces), by the frigid distinction of act and power; asserting that single bodies have each a single and proper motion, and that if they participate in any other, then this results from an external cause; and imposing countless other arbitrary restrictions on the nature of things; being always more solicitous to provide an answer to the question and affirm something positive in words, than about the inner truth of things; a failing best shown when his philosophy is compared with other systems of note among the Greeks. For the *homoeomera* of Anaxagoras; the Atoms of Leucippus and Democritus; the Heaven and Earth of Parmenides; the Strife and Friendship of Empedocles; Heraclitus' doctrine how bodies are resolved into the indifferent nature of fire, and remolded into solids, have all of them some taste of the natural philosopher—some savor of the nature of things, and experience, and bodies; whereas in the physics of Aristotle you hear hardly anything but the words of logic, which in his metaphysics also, under a more imposing name, and more forsooth as a realist than a nominalist, he has handled over again. Nor let any weight be given to the fact that in his books on animals and his problems, and other of his treatises, there is frequent dealing with experiments. For he had come to his conclusion before; he did not consult experience, as he should have done, for the purpose of framing his decisions and axioms, but having first determined the question according to his will, he then resorts to experience, and bending her into conformity with his placets, leads her about like a captive in a procession. So that even on this count he is more guilty than his modern followers, the schoolmen, who have abandoned experience altogether.

LXIV

But the Empirical school of philosophy gives birth to dogmas more deformed and monstrous than the Sophistical or Rational school. For it has its foundations not in the light of common notions (which though it be a faint and superficial light, is yet in a manner universal, and has reference to many things), but in the narrowness and darkness of a few experiments. To those therefore who are daily busied with these experiments and have infected their imagination with them, such a philosophy seems probable and all but certain; to all men else incredible and vain. Of this there is a notable instance in the alchemists and their dogmas, though it is hardly to be found elsewhere in these times, except perhaps in the philosophy of Gilbert. Nevertheless, with regard to philosophies of this kind there is one caution not to be omitted; for I foresee that if ever men are roused by my admonitions to betake themselves seriously to experiment and bid farewell to sophistical doctrines, then indeed through the premature hurry of the understanding to leap or fly to universals and principles of things, great danger may be apprehended from philosophies of this kind, against which evil we ought even now to prepare.

LXV

But the corruption of philosophy by superstition and an admixture of theology is far more widely spread, and does the greatest harm, whether to entire systems or to their parts. For the human understanding is obnoxious to the influence of the imagination no less than to the influence of common notions. For the contentious and sophistical kind of philosophy ensnares the understanding; but this kind, being fanciful and tumid and half poetical, misleads it more by flattery. For there is in man an ambition of the understanding, no less than of the will, especially in high and lofty spirits.

Of this kind we have among the Greeks a striking example in Pythagoras, though he united with it a coarser and more cumbrous superstition; another in Plato and his school, more dangerous and subtle. It shows itself likewise in parts of other philosophies, in the introduction of abstract forms and final causes and first causes, with the omission in most cases of causes intermediate, and the like. Upon this point the greatest caution should be used. For nothing is so mischievous as the apotheosis of error; and it is a very plague of the understanding for vanity to become the object of veneration. Yet in this vanity some of the moderns have with extreme levity indulged so far as to attempt to found a system of natural philosophy on the first chapter of Genesis, on the book of Job, and other parts of the sacred writings, seeking for the dead among the living; which also makes the inhibition and repression of it the more important, because from this unwholesome mixture of things human and divine there arises not only a fantastic philosophy but also a heretical religion. Very meet it is therefore that we be sober-minded, and give to faith that only which is faith's.

LXVI

So much, then, for the mischievous authorities of systems, which are founded either on common notions, or on a few experiments, or on superstition. It remains to speak of the faulty subject matter of contemplations, especially in natural philosophy. Now the human understanding is infected by the sight of what takes place in the mechanical arts, in which the alteration of bodies proceeds chiefly by composition or separation, and so imagines that something similar goes on in the universal nature of things. From this source has flowed the fiction of elements, and of their concourse for the formation of natural bodies. Again, when man contemplates nature working freely, he meets with different species of things, of animals, of plants, of minerals; whence he readily passes into the opinion that there are in nature certain primary forms which nature intends to educe, and that the remaining variety proceeds from hindrances and aberrations of nature in the fulfillment of her work, or from the collision of different species and the transplanting of one into another. To the first of these speculations we owe our primary qualities of the elements; to the other our occult properties and specific virtues; and both of them belong to those empty compendia of thought wherein the mind rests, and whereby it is diverted from more solid pursuits. It is to better purpose that the physicians bestow their labor on the secondary qualities of matter, and the operations of attraction, repulsion, attenuation, conspissation,[1] dilatation, astriction, dissipation, maturation, and the like; and were it not that by those two compendia which I have mentioned (elementary qualities, to wit, and specific virtues) they corrupted their correct observations in these other matters—either reducing them to first qualities and their subtle and incommensurable mixtures, or not following them out with greater and more diligent observations to third and fourth qualities, but breaking off the scrutiny prematurely—

---

1 [*Conspissatio.*—Ed.]

they would have made much greater progress. Nor are powers
of this kind (I do not say the same, but similar) to be sought
for only in the medicines of the human body, but also in the
changes of all other bodies.

But it is a far greater evil that they make the quiescent
principles, *wherefrom,* and not the moving principles,
whereby, things are produced, the object of their contemplation and inquiry. For the former tend to discourse, the latter to works. Nor is there any value in those vulgar distinctions of motion which are observed in the received system of
natural philosophy, as generation, corruption, augmentation,
diminution, alteration, and local motion. What they mean no
doubt is this: if a body in other respects not changed be
moved from its place, *this is local motion;* if without change
of place or essence, it be changed in quality, this is *alteration;*
if by reason of the change the mass and quantity of the body
do not remain the same, this is *augmentation* or *diminution;*
if they be changed to such a degree that they change their
very essence and substance and turn to something else, this is
generation and *corruption.* But all this is merely popular, and
does not at all go deep into nature; for these are only measures and limits, not kinds of motion. What they intimate is
how far, not *by what means,* or *from what source.* For they
do not suggest anything with regard either to the desires of
bodies or to the development of their parts. It is only when
that motion presents the thing grossly and palpably to the
sense as different from what it was that they begin to mark
the division. Even when they wish to suggest something with
regard to the causes of motion, and to establish a division
with reference to them, they introduce with the greatest
negligence a distinction between motion natural and violent,
a distinction which is itself drawn entirely from a vulgar notion, since all violent motion is also in fact natural; the external efficient simply setting nature working otherwise than
it was before. But if, leaving all this, anyone shall observe (for
instance) that there is in bodies a desire of mutual contact, so
as not to suffer the unity of nature to be quite separated or

broken and a vacuum thus made; or if anyone say that there is in bodies a desire of resuming their natural dimensions or tension, so that if compressed within or extended beyond them, they immediately strive to recover themselves, and fall back to their old volume and extent; or if anyone say that there is in bodies a desire of congregating toward masses of kindred nature—of dense bodies, for instance, toward the globe of the earth, of thin and rare bodies toward the compass of the sky; all these and the like are truly physical kinds of motion—but those others are entirely logical and scholastic, as is abundantly manifest from this comparison.

Nor again is it a lesser evil that in their philosophies and contemplations their labor is spent in investigating and handling the first principles of things and the highest generalities of nature; whereas utility and the means of working result entirely from things intermediate. Hence it is that men cease not from abstracting nature till they come to potential and uninformed matter, nor on the other hand from dissecting nature till they reach the atom; things which, even if true, can do but little for the welfare of mankind.

LXVII

A caution must also be given to the understanding against the intemperance which systems of philosophy manifest in giving or withholding assent, because intemperance of this kind seems to establish idols and in some sort to perpetuate them, leaving no way open to reach and dislodge them.

This excess is of two kinds: the first being manifest in those who are ready in deciding, and render sciences dogmatic and magisterial; the other in those who deny that we can know anything, and so introduce a wandering kind of inquiry that leads to nothing; of which kinds the former subdues, the latter weakens the understanding. For the philosophy of Aristotle, after having by hostile confutations destroyed all the rest (as the Ottomans serve their brothers), has laid down the law on all points; which done, he proceeds himself to raise new questions of his own suggestion, and dispose of them likewise, so that nothing may remain that is not certain and decided; a practice which holds and is in use among his successors.

The school of Plato, on the other hand, introduced *Acatalepsia*, at first in jest and irony, and in disdain of the older sophists, Protagoras, Hippias, and the rest, who were of nothing else so much ashamed as of seeming to doubt about anything. But the New Academy made a dogma of it, and held it as a tenet. And though theirs is a fairer seeming way than arbitrary decisions, since they say that they by no means destroy all investigation, like Pyrrho and his Refrainers, but allow of some things to be followed as probable, though of none to be maintained as true; yet still when the human mind has once despaired of finding truth, its interest in all things grows fainter, and the result is that men turn aside to pleasant disputations and discourses and roam as it were from object to object, rather than keep on a course of severe inquisition. But, as I said at the beginning and am ever urging, the human senses and understanding, weak as they are, are not to be deprived of their authority, but to be supplied with helps.

LXVIII

So much concerning the several classes of Idols and their equipage; all of which must be renounced and put away with a fixed and solemn determination, and the understanding thoroughly freed and cleansed; the entrance into the kingdom of man, founded on the sciences, being not much other than the entrance into the kingdom of heaven, whereinto none may enter except as a little child.

LXIX

But vicious demonstrations are as the strongholds and defenses of idols; and those we have in logic do little else than make the world the bondslave of human thought, and human thought the bondslave of words. Demonstrations truly are in effect the philosophies themselves and the sciences. For such as *they* are, well or ill established, such are the systems of philosophy and the contemplations which follow. Now in the whole of the process which leads from the sense and objects to axioms and conclusions, the demonstrations which we use are deceptive and incompetent. This process consists of four parts, and has as many faults. In the first place, the impressions of the sense itself are faulty; for the sense both fails us and deceives us. But its shortcomings are to be supplied, and its deceptions to be corrected. Secondly, notions are ill-drawn from the impressions of the senses, and are indefinite and confused, whereas they should be definite and distinctly bounded. Thirdly, the induction is amiss which infers the principles of sciences by simple enumeration, and does not, as it ought, employ exclusions and solutions (or separations) of nature. Lastly, that method of discovery and proof according to which the most general principles are first established, and then intermediate axioms are tried and proved by them, is the parent of error and the curse of all science. Of these things, however, which now I do but touch upon, I will speak more largely when, having performed these expiations and purgings of the mind, I come to set forth the true way for the interpretation of nature.

LXX

But the best demonstration by far is experience, if it go not beyond the actual experiment. For if it be transferred to other cases which are deemed similar, unless such transfer be made by a just and orderly process, it is a fallacious thing. But the manner of making experiments which men now use is blind and stupid. And therefore, wandering and straying as they do with no settled course, and taking counsel only from things as they fall out, they fetch a wide circuit and meet with many matters, but make little progress; and sometimes are full of hope, sometimes are distracted; and always find that there is something beyond to be sought. For it generally happens that men make their trials carelessly, and as it were in play; slightly varying experiments already known, and, if the thing does not answer, growing weary and abandoning the attempt. And even if they apply themselves to experiments more seriously and earnestly and laboriously, still they spend their labor in working out some one experiment, as Gilbert with the magnet, and the chemists with gold; a course of proceeding not less unskillful in the design than small in the attempt. For no one successfully investigates the nature of a thing in the thing itself; the inquiry must be enlarged so as to become more general.

And even when they seek to educe some science or theory from their experiments, they nevertheless almost always turn aside with overhasty and unseasonable eagerness to practice; not only for the sake of the uses and fruits of the practice, but from impatience to obtain in the shape of some new work an assurance for themselves that it is worth their while to go on; and also to show themselves off to the world, and so raise the credit of the business in which they are engaged. Thus, like Atalanta, they go aside to pick up the golden apple, but meanwhile they interrupt their course, and let the victory escape them. But in the true course of experience, and in carrying it on to the effecting of new works, the divine wisdom and order must be our pattern. Now God on the first day of creation created light only, giving to that work an entire day, in which no material substance was created. So must we likewise from experience of every kind first endeavor to discover true causes and axioms; and seek for experiments of Light, not for experiments of Fruit. For axioms rightly discovered and established supply practice with its instruments, not one by one, but in clusters, and draw after them trains and troops of works. Of the paths, however, of experience, which no less than the paths of judgment are impeded and beset, I will speak hereafter; here I have only mentioned ordinary experimental research as a bad kind of demonstration. But now the order of the matter in hand leads me to add something both as to those *signs* which I lately mentioned (signs that the systems of philosophy and contemplation in use are in a bad condition), and also as to the *causes* of what seems at first so strange and incredible. For a knowledge of the signs prepares assent; an explanation of the causes removes the marvel —which two things will do much to render the extirpation of idols from the understanding more easy and gentle.

LXXI

The sciences which we possess come for the most part from the Greeks. For what has been added by Roman, Arabic, or later writers is not much nor of much importance; and whatever it is, it is built on the foundation of Greek discoveries. Now the wisdom of the Greeks was professorial and much given to disputations, a kind of wisdom most adverse to the inquisition of truth. Thus that name of Sophists, which by those who would be thought philosophers was in contempt cast back upon and so transferred to the ancient rhetoricians, Gorgias, Protagoras, Hippias, Polus, does indeed suit the entire class: Plato, Aristotle, Zeno, Epicurus, Theophrastus, and their successors Chrysippus, Carneades, and the rest. There was this difference only, that the former class was wandering and mercenary, going about from town to town, putting up their wisdom to sale, and taking a price for it, while the latter was more pompous and dignified, as composed of men who had fixed abodes, and who opened schools and taught their philosophy without reward. Still both sorts, though in other respects unequal, were professorial; both turned the matter into disputations, and set up and battled for philosophical sects and heresies; so that their doctrines were for the most part (as Dionysius not unaptly rallied Plato) "the talk of idle old men to ignorant youths." But the elder of the Greek philosophers, Empedocles, Anaxagoras, Leucippus, Democritus, Parmenides, Heraclitus, Xenophanes, Philolaus, and the rest (I omit Pythagoras as a mystic), did not, so far as we know, open schools; but more silently and severely and simply—that is, with less affectation and parade—betook themselves to the inquisition of truth. And therefore they were in my judgment more successful; only that their works were in the course of time obscured by those slighter persons who had more which suits and pleases the capacity and tastes of the vulgar; time, like a river, bringing down to us things which are light and puffed up, but letting weighty matters sink. Still even they were not altogether free from the failing of their nation, but leaned too much to the ambition and vanity of founding a sect and catching popular applause. But the inquisition of truth must be despaired of when it turns aside to trifles of this kind. Nor should we omit that judgment, or rather divination, which was given concerning the Greeks by the Egyptian priest—that "they were always boys, without antiquity of knowledge or knowledge of antiquity." Assuredly they have that which is characteristic of boys: they are prompt to prattle, but cannot generate; for their wisdom abounds in words but is barren of works. And therefore the signs which are taken from the origin and birthplace of the received philosophy are not good.

LXXII

Nor does the character of the time and age yield much better signs than the character of the country and nation. For at that period there was but a narrow and meager knowledge either of time or place, which is the worst thing that can be, especially for those who rest all on experience. For they had no history worthy to be called history that went back a thousand years—but only fables and rumors of antiquity. And of the regions and districts of the world they knew but a small portion, giving indiscriminately the name of Scythians to all in the North, of Celts to all in the West; knowing nothing of Africa beyond the hither side of Ethiopia, of Asia beyond the Ganges. Much less were they acquainted with the provinces of the New World, even by hearsay or any well-founded rumor; nay, a multitude of climates and zones, wherein innumerable nations breathe and live, were pronounced by them to be uninhabitable; and the travels of Democritus, Plato, and Pythagoras, which were rather suburban excursions than distant journeys, were talked of as something great. In our times, on the other hand, both many parts of the New World and the limits on every side of the Old World are known, and our stock of experience has increased to an infinite amount. Wherefore if (like astrologers) we draw signs from the season of their nativity or birth, nothing great can be predicted of those systems of philosophy.

LXXIII

Of all signs there is none more certain or more noble than that taken from fruits. For fruits and works are as it were sponsors and sureties for the truth of philosophies. Now, from all these systems of the Greeks, and their ramifications through particular sciences, there can hardly after the lapse of so many years be adduced a single experiment which tends to relieve and benefit the condition of man, and which can with truth be referred to the speculations and theories of philosophy. And Celsus ingenuously and wisely owns as much when he tells us that the experimental part of medicine was first discovered, and that afterwards men philosophized about it, and hunted for and assigned causes; and not by an inverse process that philosophy and the knowledge of causes led to the discovery and development of the experimental part. And therefore it was not strange that among the Egyptians, who rewarded inventors with divine honors and sacred rites, there were more images of brutes than of men; inasmuch as brutes by their natural instinct have produced many discoveries, whereas men by discussion and the conclusions of reason have given birth to few or none.

Some little has indeed been produced by the industry of chemists; but it has been produced accidentally and in passing, or else by a kind of variation of experiments, such as mechanics use, and not by any art or theory. For the theory which they have devised rather confuses the experiments than aids them. They, too, who have busied themselves with natural magic, as they call it, have but few discoveries to show, and those trifling and imposture-like. Wherefore, as in religion we are warned to show our faith by works, so in philosophy by the same rule the system should be judged of by its fruits, and pronounced frivolous if it be barren, more especially if, in place of fruits of grape and olive, it bear thorns and briers of dispute and contention.

LXXIV

Signs also are to be drawn from the increase and progress of systems and sciences. For what is founded on nature grows and increases, while what is founded on opinion varies but increases not. If therefore those doctrines had not plainly been like a plant torn up from its roots, but had remained attached to the womb of nature and continued to draw nourishment from her, that could never have come to pass which we have seen now for twice a thousand years; namely, that the sciences stand where they did and remain almost in the same condition, receiving no noticeable increase, but on the contrary, thriving most under their first founder, and then declining. Whereas in the mechanical arts, which are founded on nature and the light of experience, we see the contrary happen, for these (as long as they are popular) are continually thriving and growing, as having in them a breath of life, at the first rude, then convenient, afterwards adorned, and at all times advancing.

LXXV

There is still another sign remaining (if sign it can be called, when it is rather testimony, nay, of all testimony the most valid). I mean the confession of the very authorities whom men now follow. For even they who lay down the law

on all things so confidently, do still in their more sober moods
fall to complaints of the subtlety of nature, the obscurity of
things, and the weakness of the human mind. Now if this
were all they did, some perhaps of a timid disposition might
be deterred from further search, while others of a more ardent
and hopeful spirit might be whetted and incited to go on
farther. But not content to speak for themselves, whatever is
beyond their own or their master's knowledge or reach they
set down as beyond the bounds of possibility, and pronounce,
as if on the authority of their art, that it cannot be known or
done; thus most presumptuously and invidiously turning the
weakness of their own discoveries into a calumny of nature
herself, and the despair of the rest of the world. Hence the
school of the New Academy, which held *Acatalepsia* as a tenet
and doomed men to perpetual darkness. Hence the opinion
that forms or true differences of things (which are in fact laws
of pure act) are past finding out and beyond the reach of man.
Hence, too, those opinions in the department of action and
operation; as, that the heat of the sun and of fire are quite different in kind—lest men should imagine that by the operations of fire anything like the works of nature can be educed
and formed. Hence the notion that composition only is the
work of man, and mixture of none but nature—lest men
should expect from art some power of generating or trans-
forming natural bodies. By this sign, therefore, men will easily
take warning not to mix up their fortunes and labors with
dogmas not only despaired of but dedicated to despair.

LXXVI

Neither is this other sign to be omitted: that formerly there
existed among philosophers such great disagreement, and such
diversities in the schools themselves, a fact which sufficiently
shows that the road from the senses to the understanding was
not skillfully laid out, when the same groundwork of philosophy (the nature of things to wit) was torn and split up
into such vague and multifarious errors. And although in

these times disagreements and diversities of opinion on first principles and entire systems are for the most part extinguished, still on parts of philosophy there remain innumerable questions and disputes, so that it plainly appears that neither in the systems themselves nor in the modes of demonstration is there anything certain or sound.

LXXVII

And as for the general opinion that in the philosophy of Aristotle, at any rate, there is great agreement, since after its publication the systems of older philosophers died away, while in the times which followed nothing better was found, so that it seems to have been so well laid and established as to have drawn both ages in its train—I answer in the first place, that the common notion of the falling off of the old systems upon the publication of Aristotle's works is a false one; for long afterwards, down even to the times of Cicero and subsequent ages, the works of the old philosophers still remained. But in the times which followed, when on the inundation of barbarians into the Roman empire human learning had suffered shipwreck, then the systems of Aristotle and Plato, like planks of lighter and less solid material, floated on the waves of time and were preserved. Upon the point of consent also men are deceived, if the matter be looked into more keenly. For true consent is that which consists in the coincidence of free judgments, after due examination. But far the greater number of those who have assented to the philosophy of Aristotle have addicted themselves thereto from prejudgment and upon the authority of others; so that it is a following and going along together, rather than consent. But even if it had been a real and widespread consent, still so little ought consent to be deemed a sure and solid confirmation, that it is in fact a strong presumption the other way. For the worst of all auguries is from consent in matters intellectual (divinity excepted, and politics where there is right of vote). For nothing pleases the many unless it strikes the imagination, or binds the understanding with the bands of common notions, as I have already said. We may very well transfer, therefore, from moral to intellectual matters the saying of Phocion, that if the multitude assent and applaud, men ought immediately to examine themselves as to what blunder or fault they may have committed. This sign, therefore, is one of the most unfavorable. And so much for this point; namely, that the signs of truth and soundness in the received systems and sciences are not good, whether they be drawn from their origin, or from their fruits, or from their progress, or from the confessions of their founders, or from general consent.

LXXVIII

I now come to the *causes* of these errors, and of so long a continuance in them through so many ages, which are very many and very potent; that all wonder how these considerations which I bring forward should have escaped men's notice till now may cease, and the only wonder be how now at last they should have entered into any man's head and become the subject of his thoughts—which truly I myself esteem as the result of some happy accident, rather than of any excellence of faculty in me—a birth of Time rather than a birth of Wit. Now, in the first place, those so many ages, if you weigh the case truly, shrink into a very small compass. For out of the five and twenty centuries over which the memory and learning of men extends, you can hardly pick out six that were fertile in sciences or favorable to their development. In times no less than in regions there are wastes and deserts. For only three revolutions and periods of learning can properly be reckoned: one among the Greeks, the second among the Romans, and the last among us, that is to say, the nations of Western Europe. And to each of these hardly two centuries can justly be assigned. The intervening ages of the world, in respect of any rich or flourishing growth of the sciences, were unprosperous. For neither the Arabians nor the Schoolmen need be mentioned, who in the intermediate times rather crushed the sciences with a multitude of treatises, than increased their weight. And therefore the first cause of so meager a progress in the sciences is duly and orderly referred to the narrow limits of the time that has been favorable to them.

LXXIX

In the second place there presents itself a cause of great weight in all ways, namely, that during those very ages in which the wits and learning of men have flourished most, or indeed flourished at all, the least part of their diligence was given to natural philosophy. Yet this very philosophy it is that ought to be esteemed the great mother of the sciences. For all arts and all sciences, if torn from this root, though they may be polished and shaped and made fit for use, yet they will hardly grow. Now it is well known that after the Christian religion was received and grew strong, by far the greater number of the best wits applied themselves to theology; that to this both the highest rewards were offered, and helps of all kinds most abundantly supplied; and that this devotion to theology chiefly occupied that third portion or epoch of time among us Europeans of the West, and the more so because about the same time both literature began to flourish and religious controversies to spring up. In the age before, on the other hand, during the continuance of the second period among the Romans, the meditations and labors of philosophers were principally employed and consumed on moral philosophy, which to the heathen was as theology to us. Moreover, in those times the greatest wits applied themselves very generally to public affairs, the magnitude of the Roman empire requiring the services of a great number of persons. Again, the age in which natural philosophy was seen to flourish most among the Greeks was but a brief particle of time; for in early ages the Seven Wise Men, as they were called (all except Thales), applied themselves to morals and politics; and in later times, when Socrates had drawn down philosophy from heaven to earth, moral philosophy became more fashionable than ever, and diverted the minds of men from the philosophy of nature.

Nay, the very period itself in which inquiries concerning nature flourished, was by controversies and the ambitious display of new opinions corrupted and made useless. Seeing therefore that during those three periods natural philosophy was in a great degree either neglected or hindered, it is no wonder if men made but small advance in that to which they were not attending.

LXXX

To this it may be added that natural philosophy, even among those who have attended to it, has scarcely ever possessed, especially in these later times, a disengaged and whole man (unless it were some monk studying in his cell, or some gentleman in his country house), but that it has been made merely a passage and bridge to something else. And so this great mother of the sciences has with strange indignity been degraded to the offices of a servant, having to attend on the business of medicine or mathematics, and likewise to wash and imbue youthful and unripe wits with a sort of first dye, in order that they may be the fitter to receive another afterwards. Meanwhile let no man look for much progress in the sciences—especially in the practical part of them—unless natural philosophy be carried on and applied to particular sciences, and particular sciences be carried back again to natural philosophy. For want of this, astronomy, optics, music, a number of mechanical arts, medicine itself—nay, what one might more wonder at, moral and political philosophy, and the logical sciences—altogether lack profoundness, and merely glide along the surface and variety of things. Because after these particular sciences have been once distributed and established, they are no more nourished by natural philosophy, which might have drawn out of the true contemplation of motions, rays, sounds, texture and configuration of bodies, affections, and intellectual perceptions, the means of imparting to them fresh strength and growth. And therefore it is nothing strange if the sciences grow not, seeing they are parted from their roots.

LXXXI

Again there is another great and powerful cause why the sciences have made but little progress, which is this. It is not possible to run a course aright when the goal itself has not been rightly placed. Now the true and lawful goal of the sciences is none other than this: that human life be endowed with new discoveries and powers. But of this the great majority have no feeling, but are merely hireling and professorial; except when it occasionally happens that some workman of acuter wit and covetous of honor applies himself to a new invention, which he mostly does at the expense of his fortunes. But in general, so far are men from proposing to themselves to augment the mass of arts and sciences, that from the mass already at hand they neither take nor look for anything more than what they may turn to use in their lectures, or to gain, or to reputation, or to some similar advantage. And if any one out of all the multitude court science with honest affection and for her own sake, yet even with him the object will be found to be rather the variety of contemplations and doctrines than the severe and rigid search after truth. And if by chance there be one who seeks after truth in earnest, yet even he will propose to himself such a kind of truth as shall yield satisfaction to the mind and understanding in rendering causes for things long since discovered, and not the truth which shall lead to new assurance of works and new light of axioms. If then the end of the sciences has not as yet been well placed, it is not strange that men have erred as to the means.

LXXXII

And as men have misplaced the end and goal of the sciences, so again, even if they had placed it right, yet they have chosen a way to it which is altogether erroneous and impass-able. And an astonishing thing it is to one who rightly con-siders the matter, that no mortal should have seriously ap-plied himself to the opening and laying out of a road for the human understanding direct from the sense, by a course of experiment orderly conducted and well built up, but that all has been left either to the mist of tradition, or the whirl and eddy of argument, or the fluctuations and mazes of chance and of vague and ill-digested experience. Now let any man soberly and diligently consider what the way is by which men have been accustomed to proceed in the investigation and dis-covery of things, and in the first place he will no doubt re-mark a method of discovery very simple and inartificial, which is the most ordinary method, and is no more than this. When a man addresses himself to discover something, he first seeks out and sets before him all that has been said about it by others; then he begins to meditate for himself; and so by much agitation and working of the wit solicits and as it were evokes his own spirit to give him oracles; which method has no foun-dation at all, but rests only upon opinions and is carried about with them.

Another may perhaps call in logic to discover it for him, but that has no relation to the matter except in name. For logical invention does not discover principles and chief ax-ioms, of which arts are composed, but only such things as ap-pear to be consistent with them. For if you grow more curious and importunate and busy, and question her of probations and invention of principles or primary axioms, her answer is well known; she refers you to the faith you are bound to give to the principles of each separate art.

There remains simple experience which, if taken as it comes, is called accident; if sought for, experiment. But this kind of experience is no better than a broom without its band, as the saying is—a mere groping, as of men in the dark, that feel all round them for the chance of finding their way, when they had much better wait for daylight, or light a candle, and then go. But the true method of experience, on the contrary, first lights the candle, and then by means of the candle shows the way; commencing as it does with experience duly ordered and digested, not bungling or erratic, and from it educing axioms, and from established axioms again new experiments; even as it was not without order and method that the divine word operated on the created mass. Let men therefore cease to wonder that the course of science is not yet wholly run, seeing that they have gone altogether astray, either leaving and abandoning experience entirely, or losing their way in it and wandering round and round as in a labyrinth. Whereas a method rightly ordered leads by an unbroken route through the woods of experience to the open ground of axioms.

LXXXIII

This evil, however, has been strangely increased by an opinion or conceit, which though of long standing is vain and hurtful, namely, that the dignity of the human mind is impaired by long and close intercourse with experiments and particulars, subject to sense and bound in matter; especially as they are laborious to search, ignoble to meditate, harsh to deliver, illiberal to practice, infinite in number, and minute in subtlety. So that it has come at length to this, that the true way is not merely deserted, but shut out and stopped up; experience being, I do not say abandoned or badly managed, but rejected with disdain.

LXXXIV

Again, men have been kept back as by a kind of enchantment from progress in the sciences by reverence for antiquity, by the authority of men accounted great in philosophy, and then by general consent. Of the last I have spoken above.

As for antiquity, the opinion touching it which men entertain is quite a negligent one and scarcely consonant with the word itself. For the old age of the world is to be accounted the true antiquity; and this is the attribute of our own times, not of that earlier age of the world in which the ancients lived, and which, though in respect of us it was the elder, yet in respect of the world it was the younger. And truly as we look for greater knowledge of human things and a riper judgment in the old man than in the young, because of his experience and of the number and variety of the things which he has seen and heard and thought of, so in like manner from our age, if it but knew its own strength and chose to essay and exert it, much more might fairly be expected than from the ancient times, inasmuch as it is a more advanced age of the world, and stored and stocked with infinite experiments and observations.

Nor must it go for nothing that by the distant voyages and travels which have become frequent in our times many things in nature have been laid open and discovered which may let in new light upon philosophy. And surely it would be disgraceful if, while the regions of the material globe—that is, of the earth, of the sea, and of the stars—have been in our times laid widely open and revealed, the intellectual globe should remain shut up within the narrow limits of old discoveries.

And with regard to authority, it shows a feeble mind to grant so much to authors and yet deny time his rights, who is the author of authors, nay, rather of all authority. For rightly is truth called the daughter of time, not of authority. It is no wonder therefore if those enchantments of antiquity and authority and consent have so bound up men's powers that they have been made impotent (like persons bewitched) to accompany with the nature of things.

LXXXV

Nor is it only the admiration of antiquity, authority, and consent, that has forced the industry of man to rest satisfied with the discoveries already made, but also an admiration for the works themselves of which the human race has long been in possession. For when a man looks at the variety and the beauty of the provision which the mechanical arts have brought together for men's use, he will certainly be more inclined to admire the wealth of man than to feel his wants; not considering that the original observations and operations of nature (which are the life and moving principle of all that variety) are not many nor deeply fetched, and that the rest is but patience, and the subtle and ruled motion of the hand and instruments—as the making of clocks (for instance) is certainly a subtle and exact work: their wheels seem to imitate the celestial orbs, and their alternating and orderly motion, the pulse of animals; and yet all this depends on one or two axioms of nature.

Again, if you observe the refinement of the liberal arts, or even that which relates to the mechanical preparation of natural substances, and take notice of such things as the discovery in astronomy of the motions of the heavens, of harmony in music, of the letters of the alphabet (to this day not in use among the Chinese) in grammar; or again in things mechanical, the discovery of the works of Bacchus and Ceres— that is, of the arts of preparing wine and beer, and of making bread; the discovery once more of the delicacies of the table, of distillations and the like; and if you likewise bear in mind the long periods which it has taken to bring these things to their present degree of perfection (for they are all ancient except distillation), and again (as has been said of clocks) how little they owe to observations and axioms of nature, and how easily and obviously and as it were by casual suggestion they may have been discovered; you will easily cease from wondering, and on the contrary will pity the condition of mankind, seeing that in a course of so many ages there has been so great a dearth and barrenness of arts and inventions. And yet these very discoveries which we have just mentioned are older than philosophy and intellectual arts. So that, if the truth must be spoken, when the rational and dogmatical sciences began, the discovery of useful works came to an end.

And again, if a man turn from the workshop to the library, and wonder at the immense variety of books he sees there, let

him but examine and diligently inspect their matter and con-
tents, and his wonder will assuredly be turned the other way.
For after observing their endless repetitions, and how men are
ever saying and doing what has been said and done before,
he will pass from admiration of the variety to astonishment at
the poverty and scantiness of the subjects which till now have
occupied and possessed the minds of men.

And if again he descend to the consideration of those arts
which are deemed curious rather than safe, and look more
closely into the works of the alchemists or the magicians, he
will be in doubt perhaps whether he ought rather to laugh
over them or to weep. For the alchemist nurses eternal hope
and when the thing fails, lays the blame upon some error of
his own; fearing either that he has not sufficiently understood
the words of his art or of his authors (whereupon he turns to
tradition and auricular whispers), or else that in his manipu-
lations he has made some slip of a scruple in weight or a moment in time (whereupon he repeats his trials to infinity). And
when, meanwhile, among the chances of experiment he lights
upon some conclusions either in aspect new or for utility not
contemptible, he takes these for earnest of what is to come,
and feeds his mind upon them, and magnifies them to the
most, and supplies the rest in hope. Not but that the alchemists
have made a good many discoveries and presented men with
useful inventions. But their case may be well compared to
the fable of the old man who bequeathed to his sons gold
buried in a vineyard, pretending not to know the exact spot;
whereupon the sons applied themselves diligently to the digging of the vineyard, and though no gold was found there,
yet the vintage by that digging was made more plentiful.

Again the students of natural magic, who explain everything by sympathies and antipathies, have in their idle and
most slothful conjectures ascribed to substances wonderful virtues and operations; and if ever they have produced works,
they have been such as aim rather at admiration and novelty
than at utility and fruit.

In superstitious magic on the other hand (if of this also we must speak), it is especially to be observed that they are but subjects of a certain and definite kind wherein the curious and superstitious arts, in all nations and ages, and religions also, have worked or played. These therefore we may pass. Meanwhile it is nowise strange if opinion of plenty has been the cause of want.

LXXXVI

Further, this admiration of men for knowledges and arts— an admiration in itself weak enough, and well-nigh childish —has been increased by the craft and artifices of those who have handled and transmitted sciences. For they set them forth with such ambition and parade, and bring them into the view of the world so fashioned and masked as if they were complete in all parts and finished. For if you look at the method of them and the divisions, they seem to embrace and comprise everything which can belong to the subject. And although these divisions are ill filled out and are but as empty cases, still to the common mind they present the form and plan of a perfect science. But the first and most ancient seekers after truth were wont, with better faith and better fortune, too, to throw the knowledge which they gathered from the contemplation of things, and which they meant to store up for use, into aphorisms; that is, into short and scattered sentences, not linked together by an artificial method; and did not pretend or profess to embrace the entire art. But as the matter now is, it is nothing strange if men do not seek to advance in things delivered to them as long since perfect and complete.

LXXXVII

Moreover, the ancient systems have received no slight accession of reputation and credit from the vanity and levity of those who have propounded new ones, especially in the active and practical department of natural philosophy. For there have not been wanting talkers and dreamers who, partly from credulity, partly in imposture, have loaded mankind with promises, offering and announcing the prolongation of life, the retardation of age, the alleviation of pain, the repairing of natural defects, the deceiving of the senses; arts of binding and inciting the affections, of illuminating and exalting the intellectual faculties, of transmuting substances, of strengthening and multiplying motions at will, of making impressions and alterations in the air, of bringing down and procuring celestial influences; arts of divining things future, and bringing things distant near, and revealing things secret; and many more. But with regard to these lavish promisers, this judgment would not be far amiss: that there is as much difference in philosophy between their vanities and true arts as there is in history between the exploits of Julius Caesar or Alexander the Great, and the exploits of Amadis of Gaul or Arthur of Britain. For it is true that those illustrious generals really did greater things than these shadowy heroes are even feigned to have done; but they did them by means and ways of action not fabulous or monstrous. Yet surely it is not fair that the credit of true history should be lessened because it has sometimes been injured and wronged by fables. Meanwhile it is not to be wondered at if a great prejudice is raised against new propositions, especially when works are also mentioned, because of those impostors who have attempted the like; since their excess of vanity, and the disgust it has bred, have their effect still in the destruction of all greatness of mind in enterprises of this kind.

LXXXVIII

Far more, however, has knowledge suffered from littleness of spirit and the smallness and slightness of the tasks which human industry has proposed to itself. And what is worst of all, this very littleness of spirit comes with a certain air of arrogance and superiority.

For in the first place there is found in all arts one general device, which has now become familiar—that the author lays the weakness of his art to the charge of nature: whatever his art cannot attain he sets down on the authority of the same art to be in nature impossible. And truly no art can be condemned if it be judge itself. Moreover, the philosophy which is now in vogue embraces and cherishes certain tenets, the purpose of which (if it be diligently examined) is to persuade men that nothing difficult, nothing by which nature may be commanded and subdued, can be expected from art or human labor; as with respect to the doctrine that the heat of the sun and of fire differ in kind, and to that other concerning mixture, has been already observed. Which things, if they be noted accurately, tend wholly to the unfair circumscription of human power, and to a deliberate and factitious despair, which not only disturbs the auguries of hope, but also cuts the sinews and spur of industry, and throws away the chances of experience itself. And all for the sake of having their art thought perfect, and for the miserable vainglory of making it believed that whatever has not yet been discovered and comprehended can never be discovered or comprehended hereafter.

And even if a man apply himself fairly to facts, and endeavor to find out something new, yet he will confine his aim and intention to the investigation and working out of some one discovery and no more; such as the nature of the magnet, the ebb and flow of the sea, the system of the heavens, and things of this kind, which seem to be in some measure secret, and have hitherto been handled without much success. Whereas it is most unskillful to investigate the nature of anything in the thing itself, seeing that the same nature which appears in some things to be latent and hidden is in others manifest and palpable; wherefore in the former it produces wonder, in the latter excites no attention; as we find it in the nature of consistency, which in wood or stone is not observed, but is passed over under the appellation of solidity without further inquiry as to why separation or solution of continuity is avoided; while in the case of bubbles, which form themselves into certain pellicles, curiously shaped into hemispheres, so that the solution of continuity is avoided for a moment, it is thought a subtle matter. In fact, what in some things is accounted a secret has in others a manifest and well-known nature, which will never be recognized as long as the experiments and thoughts of men are engaged on the former only.

But generally speaking, in mechanics old discoveries pass for new if a man does but refine or embellish them, or unite several in one, or couple them better with their use, or make the work in greater or less volume than it was before, or the like.

Thus, then, it is no wonder if inventions noble and worthy of mankind have not been brought to light, when men have been contented and delighted with such trifling and puerile tasks, and have even fancied that in them they have been endeavoring after, if not accomplishing, some great matter.

LXXXIX

Neither is it to be forgotten that in every age natural philosophy has had a troublesome and hard to deal with adversary—namely, superstition, and the blind and immoderate zeal of religion. For we see among the Greeks that those who first proposed to men's then uninitiated ears the natural causes for thunder and for storms were thereupon found guilty of impiety. Nor was much more forbearance shown by some of the ancient fathers of the Christian church to those who on most convincing grounds (such as no one in his senses would now think of contradicting) maintained that the earth was round, and of consequence asserted the existence of the antipodes.

Moreover, as things now are, to discourse of nature is made harder and more perilous by the summaries and systems of the schoolmen who, having reduced theology into regular order as well as they were able, and fashioned it into the shape of an art, ended in incorporating the contentious and thorny philosophy of Aristotle, more than was fit, with the body of religion.

To the same result, though in a different way, tend the speculations of those who have taken upon them to deduce the truth of the Christian religion from the principles of philosophers, and to confirm it by their authority, pompously solemnizing this union of the sense and faith as a lawful marriage, and entertaining men's minds with a pleasing variety of matter, but all the while disparaging things divine by mingling them with things human. Now in such mixtures of theology with philosophy only the received doctrines of philosophy are included; while new ones, albeit changes for the better, are all but expelled and exterminated.

Lastly, you will find that by the simpleness of certain divines, access to any philosophy, however pure, is well-nigh closed. Some are weakly afraid lest a deeper search into nature should transgress the permitted limits of sober-mindedness, wrongfully wresting and transferring what is said in Holy Writ against those who pry into sacred mysteries, to the hidden things of nature, which are barred by no prohibition. Others with more subtlety surmise and reflect that if second causes are unknown everything can more readily be referred to the divine hand and rod, a point in which they think religion greatly concerned—which is in fact nothing else but to seek to gratify God with a lie. Others fear from past example that movements and changes in philosophy will end in assaults on religion. And others again appear apprehensive that in the investigation of nature something may be found to subvert or at least shake the authority of religion, especially with the unlearned. But these two last fears seem to me to savor utterly of carnal wisdom; as if men in the recesses and secret thought of their hearts doubted and distrusted the strength of religion and the empire of faith over the sense, and therefore feared that the investigation of truth in nature might be dangerous to them. But if the matter be truly considered, natural philosophy is, after the word of God, at once the surest medicine against superstition and the most approved nourishment for faith, and therefore she is rightly given to religion as her most faithful handmaid, since the one displays the will of God, the other his power. For he did not err who said, "Ye err in that ye know not the Scriptures and the power of God," thus coupling and blending in an indissoluble bond information concerning his will and meditation concerning his power. Meanwhile it is not surprising if the growth of natural philosophy is checked when religion, the thing which has most power over men's minds, has by the simpleness and incautious zeal of certain persons been drawn to take part against her.

XC

Again, in the customs and institutions of schools, academies, colleges, and similar bodies destined for the abode of learned men and the cultivation of learning, everything is found adverse to the progress of science. For the lectures and exercises there are so ordered that to think or speculate on anything out of the common way can hardly occur to any man. And if one or two have the boldness to use any liberty of judgment, they must undertake the task all by themselves; they can have no advantage from the company of others. And if they can endure this also, they will find their industry and largeness of mind no slight hindrance to their fortune. For the studies of men in these places are confined and as it were imprisoned in the writings of certain authors, from whom if any man dissent he is straightway arraigned as a turbulent person and an innovator. But surely there is a great distinction between matters of state and the arts; for the danger from new motion and from new light is not the same. In matters of state a change even for the better is distrusted, because it unsettles what is established; these things resting on authority, consent, fame and opinion, not on demonstration. But arts and sciences should be like mines, where the noise of new works and further advances is heard on every side. But though the matter be so according to right reason, it is not so acted on in practice; and the points above mentioned in the administration and government of learning put a severe restraint upon the advancement of the sciences.

XCI

Nay, even if that jealousy were to cease, still it is enough to check the growth of science that efforts and labors in this field go unrewarded. For it does not rest with the same persons to cultivate sciences and to reward them. The growth of them comes from great wits; the prizes and rewards of them are in the hands of the people, or of great persons, who are but in very few cases even moderately learned. Moreover, this kind of progress is not only unrewarded with prizes and substantial benefits; it has not even the advantage of popular applause. For it is a greater matter than the generality of men can take in, and is apt to be overwhelmed and extinguished by the gales of popular opinions. And it is nothing strange if a thing not held in honor does not prosper.

XCII

But by far the greatest obstacle to the progress of science and to the undertaking of new tasks and provinces therein is found in this—that men despair and think things impossible. For wise and serious men are wont in these matters to be altogether distrustful, considering with themselves the obscurity of nature, the shortness of life, the deceitfulness of the senses, the weakness of the judgment, the difficulty of experiment, and the like; and so supposing that in the revolution of time and of the ages of the world the sciences have their ebbs and flows; that at one season they grow and flourish, at another wither and decay, yet in such sort that when they have reached a certain point and condition they can advance no further. If therefore anyone believes or promises more, they think this comes of an ungoverned and unripened mind, and that such attempts have prosperous beginnings, become difficult as they go on, and end in confusion. Now since these are thoughts which naturally present themselves to men grave and of great judgment, we must take good heed that we be not led away by our love for a most fair and excellent object to relax or diminish the severity of our judgment. We must observe diligently what encouragement dawns upon us and from what quarter, and, putting aside the lighter breezes of hope, we must thoroughly sift and examine those which promise greater steadiness and constancy. Nay, and we must take state prudence too into our counsels, whose rule is to distrust, and to take the less favorable view of human affairs. I am now therefore to speak touching hope, especially as I am not a dealer in promises, and wish neither to force nor to ensnare men's judgments, but to lead them by the hand with their good will. And though the strongest means of inspiring hope will be to bring men to particulars, especially to particulars digested and arranged in my Tables of Discovery (the subject partly of the second, but much more of the fourth part of my Instauration), since this is not merely the promise of the thing but the thing itself; nevertheless, that everything may be done with gentleness, I will proceed with my plan of preparing men's minds, of which preparation to give hope is no unimportant part. For without it the rest tends rather to make men sad (by giving them a worse and meaner opinion of things as they are than they now have, and making them more fully to feel and know the unhappiness of their own condition) than to induce any alacrity or to whet their industry in making trial. And therefore it is fit that I publish and set forth those conjectures of mine which make hope in this matter reasonable, just as Columbus did, before that wonderful voyage of his across the Atlantic, when he gave the reasons for his conviction that new lands and continents might be discovered besides those which were known before; which reasons, though rejected at first, were afterwards made good by experience, and were the causes and beginnings of great events.

XCIII

The beginning is from God: for the business which is in hand, having the character of good so strongly impressed upon it, appears manifestly to proceed from God, who is the

author of good, and the Father of Lights. Now in divine operations even the smallest beginnings lead of a certainty to their
end. And as it was said of spiritual things, "The kingdom of
God cometh not with observation," so is it in all the greater
works of Divine Providence; everything glides on smoothly
and noiselessly, and the work is fairly going on before men are
aware that it has begun. Nor should the prophecy of Daniel
be forgotten touching the last ages of the world: "Many shall
go to and fro, and knowledge shall be increased"; clearly intimating that the thorough passage of the world (which now
by so many distant voyages seems to be accomplished, or in
course of accomplishment), and the advancement of the sciences, are destined by fate, that is, by Divine Providence, to
meet in the same age.

XCIV

Next comes a consideration of the greatest importance as
an argument of hope; I mean that drawn from the errors of
past time, and of the ways hitherto trodden. For most excellent was the censure once passed upon a government that had
been unwisely administered. "That which is the worst thing
in reference to the past, ought to be regarded as best for the
future. For if you had done all that your duty demanded, and
yet your affairs were no better, you would not have even a
hope left you that further improvement is possible. But now,
when your misfortunes are owing, not to the force of circumstances, but to your own errors, you may hope that by dismissing or correcting these errors, a great change may be made
for the better." In like manner, if during so long a course of
years men had kept the true road for discovering and culti-
vating sciences, and had yet been unable to make further progress therein, bold doubtless and rash would be the opinion that
further progress is possible. But if the road itself has been mistaken, and men's labor spent on unfit objects, it follows that
the difficulty has its rise not in things themselves, which are
not in our power, but in the human understanding, and the
use and application thereof, which admits of remedy and

medicine. It will be of great use therefore to set forth what these errors are. For as many impediments as there have been in times past from this cause, so many arguments are there of hope for the time to come. And although they have been partly touched before, I think fit here also, in plain and simple words, to represent them.

XCV

Those who have handled sciences have been either men of experiment or men of dogmas. The men of experiment are like the ant, they only collect and use; the reasoners resemble spiders, who make cobwebs out of their own substance. But the bee takes a middle course: it gathers its material from the flowers of the garden and of the field, but transforms and digests it by a power of its own. Not unlike this is the true business of philosophy; for it neither relies solely or chiefly on the powers of the mind, nor does it take the matter which it gathers from natural history and mechanical experiments and lay it up in the memory whole, as it finds it, but lays it up in the understanding altered and digested. Therefore from a closer and purer league between these two faculties, the experimental and the rational (such as has never yet been made), much may be hoped.

XCVI

We have as yet no natural philosophy that is pure; all is tainted and corrupted: in Aristotle's school by logic; in Plato's by natural theology; in the second school of Platonists, such as Proclus and others, by mathematics, which ought only to give definiteness to natural philosophy, not to generate or give it birth. From a natural philosophy pure and unmixed, better things are to be expected.

XCVII

No one has yet been found so firm of mind and purpose as resolutely to compel himself to sweep away all theories and common notions, and to apply the understanding, thus made fair and even, to a fresh examination of particulars. Thus it happens that human knowledge, as we have it, is a mere medley and ill-digested mass, made up of much credulity and much accident, and also of the childish notions which we at first imbibed.

Now if anyone of ripe age, unimpaired senses, and well-purged mind, apply himself anew to experience and particulars, better hopes may be entertained of that man. In which point I promise to myself a like fortune to that of Alexander the Great, and let no man tax me with vanity till he have heard the end; for the thing which I mean tends to the putting off of all vanity. For of Alexander and his deeds Aeschines spoke thus: "Assuredly we do not live the life of mortal men; but to this end were we born, that in after ages wonders might be told of us," as if what Alexander had done seemed to him miraculous. But in the next age Titus Livius took a better and a deeper view of the matter, saying in effect that Alexander "had done no more than take courage to despise vain apprehensions." And a like judgment I suppose may be passed on myself in future ages: that I did no great things, but simply made less account of things that were accounted great. In the meanwhile, as I have already said, there is no hope except in a new birth of science; that is, in raising it regularly up from experience and building it afresh, which no one (I think) will say has yet been done or thought of.

XCVIII

Now for grounds of experience—since to experience we must come—we have as yet had either none or very weak ones; no search has been made to collect a store of particular observations sufficient either in number, or in kind, or in certainty, to inform the understanding, or in any way adequate. On the contrary, men of learning, but easy withal and idle, have taken for the construction or for the confirmation of their philosophy certain rumors and vague fames or airs of experience, and allowed to these the weight of lawful evidence. And just as if some kingdom or state were to direct its counsels and affairs not by letters and reports from ambassadors and trustworthy messengers, but by the gossip of the streets; such exactly is the system of management introduced into philosophy with relation to experience. Nothing duly investigated, nothing verified, nothing counted, weighed, or measured, is to be found in natural history; and what in observation is loose and vague, is in information deceptive and treacherous. And if anyone thinks that this is a strange thing to say, and something like an unjust complaint, seeing that Aristotle, himself so great a man, and supported by the wealth of so great a king, has composed so accurate a history of animals; and that others with greater diligence, though less pretense, have made many additions; while others, again, have compiled copious histories and descriptions of metals, plants, and fossils; it seems that he does not rightly apprehend what it is that we are now about. For a natural history which is composed for its own sake is not like one that is collected to supply the understanding with information for the building up of philosophy. They differ in many ways, but especially in this: that the former contains the variety of natural species only, and not experiments of the mechanical arts. For even as in the business of life a man's disposition and the secret workings of his mind and affections are better discovered when he is in trouble than at other times, so likewise the secrets of nature reveal themselves more readily under the vexations of art than when they go their own way. Good hopes may therefore be conceived of natural philosophy, when natural history, which is the basis and foundation of it, has been drawn up on a better plan; but not till then.

XCIX

Again, even in the great plenty of mechanical experiments, there is yet a great scarcity of those which are of most use for the information of the understanding. For the mechanic, not troubling himself with the investigation of truth, confines his attention to those things which bear upon his particular work, and will not either raise his mind or stretch out his hand for anything else. But then only will there be good ground of hope for the further advance of knowledge when there shall be received and gathered together into natural history a variety of experiments which are of no use in themselves but simply serve to discover causes and axioms, which I call *Experimenta lucifera*, experiments of *light,* to distinguish them from those which I call *fructifera*, experiments of *fruit.*

Now experiments of this kind have one admirable property and condition: they never miss or fail. For since they are applied, not for the purpose of producing any particular effect, but only of discovering the natural cause of some effect, they answer the end equally well whichever way they turn out; for they settle the question.

C

But not only is a greater abundance of experiments to be sought for and procured, and that too of a different kind from those hitherto tried; an entirely different method, order, and process for carrying on and advancing experience must also be introduced. For experience, when it wanders in its own track, is, as I have already remarked, mere groping in the dark, and confounds men rather than instructs them. But when it shall proceed in accordance with a fixed law, in regular order, and without interruption, then may better things be hoped of knowledge.

CI

But even after such a store of natural history and experience as is required for the work of the understanding, or of philosophy, shall be ready at hand, still the understanding is by no means competent to deal with it offhand and by memory alone; no more than if a man should hope by force of memory to retain and make himself master of the computation of an ephemeris. And yet hitherto more has been done in matter of invention by thinking than by writing; and experience has not yet learned her letters. Now no course of invention can be satisfactory unless it be carried on in writing. But when this is brought into use, and experience has been taught to read and write, better things may be hoped.

CII

Moreover, since there is so great a number and army of particulars, and that army so scattered and dispersed as to distract and confound the understanding, little is to be hoped for from the skirmishings and slight attacks and desultory movements of the intellect, unless all the particulars which pertain to the subject of inquiry shall, by means of Tables of Discovery, apt, well arranged, and, as it were, animate, be drawn up and marshaled; and the mind be set to work upon the helps duly prepared and digested which these tables supply.

CIII

But after this store of particulars has been set out duly and in order before our eyes, we are not to pass at once to the investigation and discovery of new particulars or works; or at any rate if we do so we must not stop there. For although I do not deny that when all the experiments of all the arts shall have been collected and digested, and brought within one man's knowledge and judgment, the mere transferring of the experiments of one art to others may lead, by means of that experience which I term literate, to the discovery of many new things of service to the life and state of man, yet it is no great matter that can be hoped from that; but from the new light of axioms, which having been educed from those particulars by a certain method and rule, shall in their turn point out the way again to new particulars, greater things may be looked for. For our road does not lie on a level, but ascends and descends; first ascending to axioms, then descending to works.

CIV

The understanding must not, however, be allowed to jump and fly from particulars to axioms remote and of almost the highest generality (such as the first principles, as they are called, of arts and things), and taking stand upon them as truths that cannot be shaken, proceed to prove and frame the middle axioms by reference to them; which has been the practice hitherto, the understanding being not only carried that way by a natural impulse, but also by the use of syllogistic demonstration trained and inured to it. But then, and then only, may we hope well of the sciences when in a just scale of ascent, and by successive steps not interrupted or broken, we rise from particulars to lesser axioms; and then to middle axioms, one above the other; and last of all to the most general. For the lowest axioms differ but slightly from bare experience, while the highest and most general (which we now have) are notional and abstract and without solidity. But the middle are the true and solid and living axioms, on which depend the affairs and fortunes of men; and above them again, last of all, those which are indeed the most general; such, I mean, as are not abstract, but of which those intermediate axioms are really limitations.

The understanding must not therefore be supplied with wings, but rather hung with weights, to keep it from leaping and flying. Now this has never yet been done; when it is done, we may entertain better hopes of the sciences.

CV

In establishing axioms, another form of induction must be devised than has hitherto been employed, and it must be used for proving and discovering not first principles (as they are called) only, but also the lesser axioms, and the middle, and indeed all. For the induction which proceeds by simple enumeration is childish; its conclusions are precarious and exposed to peril from a contradictory instance; and it generally decides on too small a number of facts, and on those only which are at hand. But the induction which is to be available for the discovery and demonstration of sciences and arts, must analyze nature by proper rejections and exclusions; and then, after a sufficient number of negatives, come to a conclusion on the affirmative instances—which has not yet been done or even attempted, save only by Plato, who does indeed employ this form of induction to a certain extent for the purpose of discussing definitions and ideas. But in order to furnish this induction or demonstration well and duly for its work, very many things are to be provided which no mortal has yet thought of; insomuch that greater labor will have to be spent in it than has hitherto been spent on the syllogism. And this induction must be used not only to discover axioms, but also in the formation of notions. And it is in this induction that our chief hope lies.

CVI

But in establishing axioms by this kind of induction, we must also examine and try whether the axiom so established be framed to the measure of those particulars only from which it is derived, or whether it be larger and wider. And if it be larger and wider, we must observe whether by indicating to us new particulars it confirm that wideness and largeness as by a collateral security, that we may not either stick fast in things already known, or loosely grasp at shadows and abstract forms, not at things solid and realized in matter. And when this process shall have come into use, then at last shall we see the dawn of a solid hope.

CVII

And here also should be remembered what was said above concerning the extending of the range of natural philosophy to take in the particular sciences, and the referring or bringing back of the particular sciences to natural philosophy, that the branches of knowledge may not be severed and cut off from the stem. For without this the hope of progress will not be so good.

CVIII

So much then for the removing of despair and the raising of hope through the dismissal or rectification of the errors of past time. We must now see what else there is to ground hope upon. And this consideration occurs at once—that if many useful discoveries have been made by accident or upon occasion, when men were not seeking for them but were busy about other things, no one can doubt but that when they apply themselves to seek and make this their business, and that too by method and in order and not by desultory impulses, they will discover far more. For although it may happen once or twice that a man shall stumble on a thing by accident which, when taking great pains to search for it, he could not find, yet upon the whole it unquestionably falls out the other way. And therefore far better things, and more of them, and at shorter intervals, are to be expected from man's reason and industry and direction and fixed application than from accident and animal instinct and the like, in which inventions have hitherto had their origin.

CIX

Another argument of hope may be drawn from this—that some of the inventions already known are such as before they were discovered it could hardly have entered any man's head to think of; they would have been simply set aside as impossible. For in conjecturing what may be men set before them the example of what has been, and divine of the new with an imagination preoccupied and colored by the old; which way of forming opinions is very fallacious, for streams that are drawn from the springheads of nature do not always run in the old channels.

If, for instance, before the invention of ordnance, a man had described the thing by its effects, and said that there was a new invention by means of which the strongest towers and walls could be shaken and thrown down at a great distance, men would doubtless have begun to think over all the ways of multiplying the force of catapults and mechanical engines by weights and wheels and such machinery for ramming and projecting; but the notion of a fiery blast suddenly and violently expanding and exploding would hardly have entered into any man's imagination or fancy, being a thing to which nothing immediately analogous had been seen, except perhaps in an earthquake or in lightning, which as *magnalia* or marvels of nature, and by man not imitable, would have been immediately rejected.

In the same way, if, before the discovery of silk, anyone had said that there was a kind of thread discovered for the purposes of dress and furniture which far surpassed the thread of linen or of wool in fineness and at the same time in strength, and also in beauty and softness, men would have begun immediately to think of some silky kind of vegetable, or of the finer hair of some animal, or of the feathers and down of birds; but a web woven by a tiny worm, and that in such abundance, and renewing itself yearly, they would assuredly never have thought. Nay, if anyone had said anything about a worm, he would no doubt have been laughed at as dreaming of a new kind of cobwebs.

So again, if, before the discovery of the magnet, anyone had said that a certain instrument had been invented by means of which the quarters and points of the heavens could be taken and distinguished with exactness, men would have been carried by their imagination to a variety of conjectures concerning the more exquisite construction of astronomical instruments; but that anything could be discovered agreeing so well in its movements with the heavenly bodies, and yet not a heavenly body itself, but simply a substance of metal or stone, would have been judged altogether incredible. Yet these things and others like them lay for so many ages of the world concealed from men, nor was it by philosophy or the rational arts that they were found out at last, but by accident and occasion, being indeed, as I said, altogether different in kind and as remote as possible from anything that was known before; so that no preconceived notion could possibly have led to the discovery of them.

There is therefore much ground for hoping that there are still laid up in the womb of nature many secrets of excellent use, having no affinity or parallelism with anything that is now known, but lying entirely out of the beat of the imagination, which have not yet been found out. They too no doubt will some time or other, in the course and revolution of many ages, come to light of themselves, just as the others did; only by the method of which we are now treating they can be speedily and suddenly and simultaneously presented and anticipated.

CX

But we have also discoveries to show of another kind, which prove that noble inventions may be lying at our very feet, and yet mankind may step over without seeing them. For however the discovery of gunpowder, of silk, of the magnet, of sugar, of paper, or the like, may seem to depend on certain properties of things themselves and nature, there is at any rate nothing in the art of printing which is not plain and obvious. Nevertheless for want of observing that although it is more difficult to arrange types of letters than to write letters by the motion of the hand, there is yet this difference between the two, that types once arranged serve for innumerable impressions, but letters written with the hand for a single copy only; or perhaps again for want of observing that ink can be so thickened as to color without running (particularly when the letters face upwards and the impression is made from above)—for want, I say, of observing these things, men went for so many ages without this most beautiful discovery, which is of so much service in the propagation of knowledge.

But such is the infelicity and unhappy disposition of the human mind in this course of invention, that it first distrusts and then despises itself: first will not believe that any such thing can be found out; and when it is found out, cannot understand how the world should have missed it so long. And this very thing may be justly taken as an argument of hope, namely, that there is a great mass of inventions still remaining which not only by means of operations that are yet to be discovered, but also through the transferring, comparing, and applying of those already known, by the help of that learned experience of which I spoke, may be deduced and brought to light.

CXI

There is another ground of hope that must not be omitted. Let men but think over their infinite expenditure of understanding, time, and means on matters and pursuits of far less use and value; whereof, if but a small part were directed to sound and solid studies, there is no difficulty that might not be overcome. This I thought good to add, because I plainly confess that a collection of history natural and experimental, such as I conceive it and as it ought to be, is a great, I may say a royal work, and of much labor and expense.

CXII

Meantime, let no man be alarmed at the multitude of particulars, but let this rather encourage him to hope. For the particular phenomena of art and nature are but a handful to the inventions of the wit, when disjoined and separated from the evidence of things. Moreover, this road has an issue in the open ground and not far off; the other has no issue at all, but endless entanglement. For men hitherto have made but short stay with experience, but passing her lightly by, have wasted an infinity of time on meditations and glosses of the wit. But if someone were by that could answer our questions and tell us in each case what the fact in nature is, the discovery of all causes and sciences would be but the work of a few years.

CXIII

Moreover, I think that men may take some hope from my own example. And this I say not by way of boasting, but because it is useful to say it. If there be any that despond, let them look at me, that being of all men of my time the most busied in affairs of state, and a man of health not very strong (whereby much time is lost), and in this course altogether a pioneer, following in no man's track nor sharing these counsels with anyone, have nevertheless by resolutely entering on the true road, and submitting my mind to Things, advanced these matters, as I suppose, some little way. And then let them consider what may be expected (after the way has been thus indicated) from men abounding in leisure, and from association of labors, and from successions of ages—the rather because it is not a way over which only one man can pass at a time (as is the case with that of reasoning), but one in which the labors and industries of men (especially as regards the collecting of experience) may with the best effect be first distributed and then combined. For then only will men begin to know their strength when instead of great numbers doing all the same things, one shall take charge of one thing and another of another.

CXIV

Lastly, even if the breath of hope which blows on us from that New Continent were fainter than it is and harder to perceive, yet the trial (if we would not bear a spirit altogether abject) must by all means be made. For there is no comparison between that which we may lose by not trying and by not succeeding, since by not trying we throw away the chance of an immense good; by not succeeding we only incur the loss of a little human labor. But as it is, it appears to me from what has been said, and also from what has been left unsaid, that there is hope enough and to spare, not only to make a bold man try, but also to make a sober-minded and wise man believe.

CXV

Concerning the grounds then for putting away despair, which has been one of the most powerful causes of delay and hindrance to the progress of knowledge, I have now spoken. And this also concludes what I had to say touching the signs and causes of the errors, sluggishness, and ignorance which have prevailed; especially since the more subtle causes, which do not fall under popular judgment and observation, must be referred to what has been said on the Idols of the human mind.

And here likewise should close that part of my Instauration which is devoted to pulling down, which part is performed by three refutations: first, by the refutation of the *natural human reason*, left to itself; secondly, by the refutation of the *demonstrations;* and thirdly, by the refutation of the *theories,* or the received systems of philosophy and doctrine. And the refutation of these has been such as alone it could be: that is to say, by signs and the evidence of causes, since no other kind of confutation was open to me, differing as I do from the others both on first principles and on rules of demonstration.

It is time therefore to proceed to the art itself and rule of interpreting nature. Still, however, there remains something to be premised. For whereas in this first book of aphorisms I proposed to prepare men's minds as well for understanding as for receiving what is to follow, now that I have purged and swept and leveled the floor of the mind, it remains that I place the mind in a good position and as it were in a favorable aspect toward what I have to lay before it. For in a new matter it is not only the strong preoccupation of some old opinion that tends to create a prejudice, but also a false preconception or prefiguration of the new thing which is presented. I will endeavor therefore to impart sound and true opinions as to the things I propose, although they are to serve only for the time, and by way of interest (so to speak), till the thing itself, which is the principal, be fully known.

CXVI

First, then, I must request men not to suppose that after the fashion of ancient Greeks, and of certain moderns, as Telesius, Patricius, Severinus, I wish to found a new sect in philosophy. For this is not what I am about, nor do I think that it matters much to the fortunes of men what abstract notions one may entertain concerning nature and the principles of things. And no doubt many old theories of this kind can be revived and many new ones introduced, just as many theories of the heavens may be supposed which agree well enough with the phenomena and yet differ with each other.

But for my part I do not trouble myself with any such speculative and withal unprofitable matters. My purpose, on the contrary, is to try whether I cannot in very fact lay more firmly the foundations and extend more widely the limits of the power and greatness of man. And although on some special subjects and in an incomplete form I am in possession of results which I take to be far more true and more certain and withal more fruitful than those now received (and these I have collected into the fifth part of my Instauration), yet I have no entire or universal theory to propound. For it does not seem that the time is come for such an attempt. Neither can I hope to live to complete the sixth part of the Instauration (which is destined for the philosophy discovered by the legitimate interpretation of nature), but hold it enough if in the intermediate business I bear myself soberly and profitably, sowing in the meantime for future ages the seeds of a purer truth, and performing my part toward the commencement of the great undertaking.

CXVII

And as I do not seek to found a school, so neither do I hold out offers or promises of particular works. It may be thought, indeed, that I who make such frequent mention of works and refer everything to that end, should produce some myself by way of earnest. But my course and method, as I have often clearly stated and would wish to state again, is this—not to extract works from works or experiments from experiments (as an empiric), but from works and experiments to extract causes and axioms, and again from those causes and axioms new works and experiments, as a legitimate interpreter of nature. And although in my tables of discovery (which compose the fourth part of the Instauration), and also in the examples of particulars (which I have adduced in the second part), and moreover in my observations on the history (which I have drawn out in the third part), any reader of even moderate sagacity and intelligence will everywhere observe indications and outlines of many noble works; still I candidly confess that the natural history which I now have, whether collected from books or from my own investigations, is neither sufficiently copious nor verified with sufficient accuracy to serve the purposes of legitimate interpretation.

Accordingly, if there be anyone more apt and better prepared for mechanical pursuits, and sagacious in hunting out works by the mere dealing with experiment, let him by all means use his industry to gather from my history and tables many things by the way, and apply them to the production of works, which may serve as interest until the principal be forthcoming. But for myself, aiming as I do at greater things, I condemn all unseasonable and premature tarrying over such things as these, being (as I often say) like Atalanta's balls. For I do not run off like a child after golden apples, but stake all on the victory of art over nature in the race. Nor do I make haste to mow down the moss or the corn in blade, but wait for the harvest in its due season.

CXVIII

There will be found, no doubt, when my history and tables of discovery are read, some things in the experiments themselves that are not quite certain, or perhaps that are quite false, which may make a man think that the foundations and principles upon which my discoveries rest are false and doubtful. But this is of no consequence, for such things must needs happen at first. It is only like the occurrence in a written or printed page of a letter or two mistaken or misplaced, which does not much hinder the reader, because such errors are easily corrected by the sense. So likewise may there occur in my natural history many experiments which are mistaken and falsely set down, and yet they will presently, by the discovery of causes and axioms, be easily expunged and rejected. It is nevertheless true that if the mistakes in natural history and experiments are important, frequent, and continual, they cannot possibly be corrected or amended by any felicity of wit or art. And therefore, if in my natural history, which has been collected and tested with so much diligence, severity, and I may say religious care, there still lurk at intervals certain falsities or errors in the particulars, what is to be said of common natural history, which in comparison with mine is so negligent and inexact? And what of the philosophy and sciences built on such a sand (or rather quicksand)? Let no man therefore trouble himself for this.

CXIX

There will be met with also in my history and experiments many things which are trivial and commonly known; many which are mean and low; many, lastly, which are too subtle and merely speculative, and that seem to be of no use; which kind of things may possibly avert and alienate men's interest.

And first, for those things which seem common. Let men bear in mind that hitherto they have been accustomed to do no more than refer and adapt the causes of things which rarely happen to such as happen frequently, while of those which happen frequently they never ask the cause, but take them as they are for granted. And therefore they do not investigate the causes of weight, of the rotation of heavenly bodies, of heat, cold, light, hardness, softness, rarity, density, liquidity, solidity, animation, inanimation, similarity, dissimilarity, organization, and the like; but admitting these as self-evident and obvious, they dispute and decide on other things of less frequent and familiar occurrence.

But I, who am well aware that no judgment can be passed on uncommon or remarkable things, much less anything new brought to light, unless the causes of common things, and the causes of those causes, be first duly examined and found out, am of necessity compelled to admit the commonest things into my history. Nay, in my judgment philosophy has been hindered by nothing more than this, that things of familiar and frequent occurrence do not arrest and detain the thoughts of men, but are received in passing without any inquiry into their causes; insomuch that information concerning things which are not known is not oftener wanted than attention concerning things which are.

CXX

And for things that are mean or even filthy—things which (as Pliny says) must be introduced with an apology—such things, no less than the most splendid and costly, must be admitted into natural history. Nor is natural history polluted thereby, for the sun enters the sewer no less than the palace, yet takes no pollution. And for myself, I am not raising a capitol or pyramid to the pride of man, but laying a foundation in the human understanding for a holy temple after the model of the world. That model therefore I follow. For whatever deserves to exist deserves also to be known, for knowledge is the image of existence; and things mean and splendid exist alike. Moreover, as from certain putrid substances—musk, for instance, and civet—the sweetest odors are sometimes generated, so, too, from mean and sordid instances there sometimes emanates excellent light and information. But enough and more than enough of this, such fastidiousness being merely childish and effeminate.

CXXI

But there is another objection which must be more carefully looked to, namely, that there are many things in this History which to common apprehension, or indeed to any understanding accustomed to the present system, will seem to be curiously and unprofitably subtle. Upon this point, therefore, above all I must say again what I have said already: that at first, and for a time, I am seeking for experiments of light, not for experiments of fruit, following therein, as I have often said, the example of the divine creation which on the first day produced light only, and assigned to it alone one entire day, nor mixed up with it on that day any material work.

To suppose, therefore, that things like these are of no use is the same as to suppose that light is of no use, because it is not a thing solid or material. And the truth is that the knowledge of simple natures well examined and defined is as light: it gives entrance to all the secrets of nature's workshop, and virtually includes and draws after it whole bands and troops of works, and opens to us the sources of the noblest axioms; and yet in itself it is of no great use. So also the letters of the alphabet in themselves and apart have no use or meaning, yet they are the subject matter for the composition and apparatus of all discourse. So again the seeds of things are of much latent virtue, and yet of no use except in their development. And the scattered rays of light itself, until they are made to converge, can impart none of their benefit.

But if objection be taken to speculative subtleties, what is to be said of the schoolmen, who have indulged in subtleties to such excess—in subtleties, too, that were spent on words, or at any rate on popular notions (which is much the same thing), not on facts or nature; and such as were useless not only in their origin but also in their consequences; and not like those I speak of, useless indeed for the present, but promising infinite utility hereafter. But let men be assured of this, that all subtlety of disputation and discourse, if not applied till after axioms are discovered, is out of season and preposterous, and that the true and proper or at any rate the chief time for subtlety is in weighing experience and in founding axioms thereon. For that other subtlety, though it grasps and snatches at nature, yet can never take hold of her. Certainly what is said of opportunity or fortune is most true of nature: she has a lock in front, but is bald behind.

Lastly, concerning the disdain to receive into natural history things either common, or mean, or oversubtle and in their original condition useless, the answer of the poor woman to the haughty prince who had rejected her petition as an unworthy thing and beneath his dignity, may be taken for an oracle: "Then leave off being king." For most certain it is that he who will not attend to things like these as being too paltry and minute, can neither win the kingdom of nature nor govern it.

CXXII

It may be thought also a strange and a harsh thing that we should at once and with one blow set aside all sciences and all authors; and that, too, without calling in any of the ancients to our aid and support, but relying on our own strength.

And I know that if I had chosen to deal less sincerely, I might easily have found authority for my suggestions by referring them either to the old times before the Greeks (when natural science was perhaps more flourishing, though it made less noise, not having yet passed into the pipes and trumpets of the Greeks), or even, in part at least, to some of the Greeks themselves; and so gained for them both support and honor, as men of no family devise for themselves by the good help of genealogies the nobility of a descent from some ancient stock. But for my part, relying on the evidence and truth of things, I reject all forms of fiction and imposture; nor do I think that it matters any more to the business in hand whether the discoveries that shall now be made were long ago known to the ancients, and have their settings and their risings according to the vicissitude of things and course of ages, than it matters to mankind whether the new world be that island of Atlantis with which the ancients were acquainted, or now discovered for the first time. For new discoveries must be sought from the light of nature, not fetched back out of the darkness of antiquity.

And as for the universality of the censure, certainly if the matter be truly considered such a censure is not only more probable but more modest, too, than a partial one would be. For if the errors had not been rooted in primary notions, there must have been some true discoveries to correct the false. But the errors being fundamental, and not so much of false judgment as of inattention and oversight, it is no wonder that men have not obtained what they have not tried for, nor reached a mark which they never set up, nor finished a course which they never entered on or kept.

And as for the presumption implied in it, certainly if a man undertakes by steadiness of hand and power of eye to describe a straighter line or more perfect circle than anyone else, he challenges a comparison of abilities; but if he only says that he with the help of a rule or a pair of compasses can draw a straighter line or a more perfect circle than anyone else can by eye and hand alone, he makes no great boast. And this remark, be it observed, applies not merely to this first and inceptive attempt of mine, but to all that shall take the work in hand hereafter. For my way of discovering sciences goes far to level men's wit and leaves but little to individual excellence, because it performs everything by the surest rules and demonstrations. And therefore I attribute my part in all this, as I have often said, rather to good luck than to ability, and account it a birth of time rather than of wit. For certainly chance has something to do with men's thoughts, as well as with their works and deeds.

CXXIII

I may say then of myself that which one said in jest (since it marks the distinction so truly), "It cannot be that we should think alike, when one drinks water and the other drinks wine." Now other men, as well in ancient as in modern times, have in the matter of sciences drunk a crude liquor like water, either flowing spontaneously from the understanding, or drawn up by logic, as by wheels from a well. Whereas I pledge mankind in a liquor strained from countless grapes, from grapes ripe and fully seasoned, collected in clusters, and gathered, and then squeezed in the press, and finally purified and clarified in the vat. And therefore it is no wonder if they and I do not think alike.

CXXIV

Again, it will be thought, no doubt, that the goal and mark of knowledge which I myself set up (the very point which I object to in others) is not the true or the best, for that the contemplation of truth is a thing worthier and loftier than all utility and magnitude of works; and that this long and anxious dwelling with experience and matter and the fluctuations of individual things, drags down the mind to earth, or rather sinks it to a very Tartarus of turmoil and confusion, removing and withdrawing it from the serene tranquility of abstract wisdom, a condition far more heavenly. Now to this I readily assent, and indeed this which they point at as so much to be preferred is the very thing of all others which I am about. For I am building in the human understanding a true model of the world, such as it is in fact, not such as a man's own reason would have it to be; a thing which cannot be done without a very diligent dissection and anatomy of the world. But I say that those foolish and apish images of worlds which the fancies of men have created in philosophical systems must be utterly scattered to the winds. Be it known then how vast a difference there is (as I said above) between the idols of the human mind and the ideas of the divine. The former are nothing more than arbitrary abstractions; the latter are the Creator's own stamp upon creation, impressed and defined in matter by true and exquisite lines. Truth, therefore, and utility are here the very same things; [2] and works themselves are of greater value as pledges of truth than as contributing to the comforts of life.

CXXV

It may be thought again that I am but doing what has been done before; that the ancients themselves took the same course which I am now taking; and that it is likely therefore that I too, after all this stir and striving, shall come at last to some one of those systems which prevailed in ancient times. For the ancients, too, it will be said, provided at the outset of their speculations a great store and abundance of examples and particulars, digested the same into notebooks under heads and titles, from them completed their systems and arts, and afterward, when they understood the matter, published them to the world, adding a few examples here and there for proof and illustration; but thought it superfluous and inconvenient to publish their notes and minutes and digests of particulars, and therefore did as builders do: after the house was built they removed the scaffolding and ladders out of sight. And so no doubt they did. But this objection (or scruple rather) will be easily answered by anyone who has not quite forgotten what I have said above. For the form of inquiry and discovery that was in use among the ancients is by themselves professed and appears on the very face of their writings. And that form was simply this. From a few examples and particulars (with the addition of common notions and perhaps of some portion of the received opinions which have been most popular) they flew at once to the most general conclusions, or first principles of science. Taking the truth of these as fixed and immovable, they

[2] *Ipsissimæ res.* I think this must have been Bacon's meaning, though not a meaning which the word can properly bear.—*J. S.*

proceeded by means of intermediate propositions to educe and prove from them the inferior conclusions; and out of these they framed the art. After that, if any new particulars and examples repugnant to their dogmas were mooted and adduced, either they subtly molded them into their system by distinctions or explanations of their rules, or else coarsely got rid of them by exceptions; while to such particulars as were not repugnant they labored to assign causes in conformity with those of their principles. But this was not the natural history and experience that was wanted; far from it. And besides, that flying off to the highest generalities ruined all.

CXXVI

It will also be thought that by forbidding men to pronounce and to set down principles as established until they have duly arrived through the intermediate steps at the highest generalities, I maintain a sort of suspension of the judgment, and bring it to what the Greeks call *Acatalepsia*—a denial of the capacity of the mind to comprehend truth. But in reality that which I meditate and propound is not *Acatalepsia,* but *Eucatalepsia;* not denial of the capacity to understand, but provision for understanding truly. For I do not take away authority from the senses, but supply them with helps; I do not slight the understanding, but govern it. And better surely it is that we should know all we need to know, and yet think our knowledge imperfect, than that we should think our knowledge perfect, and yet not know anything we need to know.

CXXVII

It may also be asked (in the way of doubt rather than objection) whether I speak of natural philosophy only, or whether I mean that the other sciences, logic, ethics, and politics, should be carried on by this method. Now I certainly mean what I have said to be understood of them all; and as the common logic, which governs by the syllogism, extends not only to natural but to all sciences, so does mine also, which proceeds by induction, embrace everything. For I form a history and table of discovery for anger, fear, shame, and the like; for matters political; and again for the mental operations of memory, composition and division, judgment, and the rest; not less than for heat and cold, or light, or vegetation, or the like. But, nevertheless, since my method of interpretation, after the history has been prepared and duly arranged, regards not the working and discourse of the mind only (as the common logic does) but the nature of things also, I supply the mind such rules and guidance that it may in every case apply itself aptly to the nature of things. And therefore I deliver many and diverse precepts in the doctrine of interpretation, which in some measure modify the method of invention according to the quality and condition of the subject of the inquiry.

CXXVIII

On one point not even a doubt ought to be entertained, namely, whether I desire to pull down and destroy the philosophy and arts and sciences which are at present in use. So far from that, I am most glad to see them used, cultivated, and honored. There is no reason why the arts which are now in fashion should not continue to supply matter for disputation and ornaments for discourse, to be employed for the convenience of professors and men of business, to be, in short, like current coin, which passes among men by consent. Nay, I frankly declare that what I am introducing will be but little fitted for such purposes as these, since it cannot be brought down to common apprehension save by effects and works only. But how sincere I am in my professions of affection and good will toward the received sciences, my published writings, especially the books on the advancement of learning, sufficiently show; and therefore I will not attempt to prove it further by words. Meanwhile I give constant and distinct warning that by the methods now in use neither can any great progress be made in the doctrines and contemplative part of sciences, nor can they be carried out to any magnitude of works.

CXXIX

It remains for me to say a few words touching the excellency of the end in view. Had they been uttered earlier, they might have seemed like idle wishes, but now that hopes have been raised and unfair prejudices removed, they may perhaps have greater weight. Also if I had finished all myself, and had no occasion to call in others to help and take part in the work, I should even now have abstained from such language lest it might be taken as a proclamation of my own deserts. But since I want to quicken the industry and rouse and kindle the zeal of others, it is fitting that I put men in mind of some things.

In the first place, then, the introduction of famous discoveries appears to hold by far the first place among human actions; and this was the judgment of the former ages. For to the authors of inventions they awarded divine honors, while to those who did good service in the state (such as founders of cities and empires, legislators, saviors of their country from long endured evils, quellers of tyrannies, and the like) they decreed no higher honors than heroic. And certainly if a man rightly compare the two, he will find that this judgment of antiquity was just. For the benefits of discoveries may extend to the whole race of man, civil benefits only to particular places; the latter last not beyond a few ages, the former through all time. Moreover, the reformation of a state in civil matters is seldom brought in without violence and confusion; but discoveries carry blessings with them, and confer benefits without causing harm or sorrow to any.

Again, discoveries are as it were new creations, and imitations of God's works, as the poet well sang:

> To man's frail race great Athens long ago
> First gave the seed whence waving harvests grow,
> And *re-created* all our life below.

And it appears worthy of remark in Solomon that, though mighty in empire and in gold, in the magnificence of his works, his court, his household, and his fleet, in the luster of his name and the worship of mankind, yet he took none of these to glory in, but pronounced that "The glory of God is to conceal a thing; the glory of the king to search it out."

Again, let a man only consider what a difference there is between the life of men in the most civilized province of Europe, and in the wildest and most barbarous districts of New India; he will feel it be great enough to justify the saying that "man is a god to man," not only in regard to aid and benefit, but also by a comparison of condition. And this difference comes not from soil, not from climate, not from race, but from the arts.

Again, it is well to observe the force and virtue and consequences of discoveries, and these are to be seen nowhere more conspicuously than in those three which were unknown to the ancients, and of which the origin, though recent, is obscure and inglorious; namely, printing, gunpowder, and the magnet. For these three have changed the whole face and state of things throughout the world; the first in literature, the second in warfare, the third in navigation; whence have followed innumerable changes, insomuch that no empire, no sect, no star seems to have exerted greater power and influence in human affairs than these mechanical discoveries.

Further, it will not be amiss to distinguish the three kinds and, as it were, grades of ambition in mankind. The first is of those who desire to extend their own power in their native country, a vulgar and degenerate kind. The second is of those who labor to extend the power and dominion of their country among men. This certainly has more dignity, though not less covetousness. But if a man endeavor to establish and extend the power and dominion of the human race itself over the universe, his ambition (if ambition it can be called) is without doubt both a more wholesome and a more noble thing than the other two. Now the empire of man over things depends wholly on the arts and sciences. For we cannot command nature except by obeying her.

Again, if men have thought so much of some one particular discovery as to regard him as more than man who has been able by some benefit to make the whole human race his debtor, how much higher a thing to discover that by means of which all things else shall be discovered with ease! And yet (to speak the whole truth), as the uses of light are infinite in enabling us to walk, to ply our arts, to read, to recognize one another—and nevertheless the very beholding of the light is itself a more excellent and a fairer thing than all the uses of it—so assuredly the very contemplation of things as they are, without superstition or imposture, error or confusion, is in itself more worthy than all the fruit of inventions.

Lastly, if the debasement of arts and sciences to purposes of wickedness, luxury, and the like, be made a ground of objection, let no one be moved thereby. For the same may be said of all earthly goods: of wit, courage, strength, beauty, wealth, light itself, and the rest. Only let the human race recover that right over nature which belongs to it by divine bequest, and let power be given it; the exercise thereof will be governed by sound reason and true religion.

CXXX

And now it is time for me to propound the art itself of interpreting nature, in which, although I conceive that I have given true and most useful precepts, yet I do not say either that it is absolutely necessary (as if nothing could be done without it) or that it is perfect. For I am of the opinion that if men had ready at hand a just history of nature and experience, and labored diligently thereon, and if they could bind themselves to two rules—the first, to lay aside received opinions and notions; and the second, to refrain the mind for a time from the highest generalizations, and those next to them —they would be able by the native and genuine force of the

mind, without any other art, to fall into my form of interpretation. For interpretation is the true and natural work of
the mind when freed from impediments. It is true, however,
that by my precepts everything will be in more readiness, and
much more sure.

Nor again do I mean to say that no improvement can be
made upon these. On the contrary, I regard that the mind, not
only in its own faculties, but in its connection with things,
must needs hold that the art of discovery may advance as
discoveries advance.

APHORISMS

[BOOK TWO]

I

On a given body, to generate and superinduce a new nature or new natures is the work and aim of human power. Of a given nature to discover the form, or true specific difference, or nature-engendering nature, or source of emanation (for these are the terms which come nearest to a description of the thing), is the work and aim of human knowledge. Subordinate to these primary works are two others that are secondary and of inferior mark: to the former, the transformation of concrete bodies, so far as this is possible; to the latter, the discovery, in every case of generation and motion, of the *latent process* carried on from the manifest efficient and the manifest material to the form which is engendered; and in like manner the discovery of the *latent configuration* of bodies at rest and not in motion.

II

In what an ill condition human knowledge is at the present time is apparent even from the commonly received maxims. It is a correct position that "true knowledge is knowledge by causes." And causes again are not improperly distributed into four kinds: the material, the formal, the efficient, and the final. But of these the final cause rather corrupts than advances the sciences, except such as have to do with human action. The discovery of the formal is despaired of. The efficient and the material (as they are investigated and received, that is, as remote causes, without reference to the latent process leading to the form) are but slight and superficial, and contribute little, if anything, to true and active science. Nor

121

have I forgotten that in a former passage I noted and corrected as an error of the human mind the opinion that forms give existence. For though in nature nothing really exists besides individual bodies, performing pure individual acts according to a fixed law, yet in philosophy this very law, and the investigation, discovery, and explanation of it, is the foundation as well of knowledge as of operation. And it is this law with its clauses that I mean when I speak of *forms*, a name which I the rather adopt because it has grown into use and become familiar.

III

If a man be acquainted with the cause of any nature (as whiteness or heat) in certain subjects only, his knowledge is imperfect; and if he be able to superinduce an effect on certain substances only (of those susceptible of such effect), his power is in like manner imperfect. Now if a man's knowledge be confined to the efficient and material causes (which are unstable causes, and merely vehicles, or causes which convey the form in certain cases) he may arrive at new discoveries in reference to substances in some degree similar to one another, and selected beforehand; but he does not touch the deeper boundaries of things. But whosoever is acquainted with forms embraces the unity of nature in substances the most unlike, and is able therefore to detect and bring to light things never yet done, and such as neither the vicissitudes of nature, nor industry in experimenting, nor accident itself, would ever have brought into act, and which would never have occurred to the thought of man. From the discovery of forms therefore results truth in speculation and freedom in operation.

IV

Although the roads to human power and to human knowledge lie close together and are nearly the same, nevertheless, on account of the pernicious and inveterate habit of dwelling on abstractions it is safer to begin and raise the sciences from those foundations which have relation to practice, and to let the active part itself be as the seal which prints and determines the contemplative counterpart. We must therefore consider, if a man wanted to generate and superinduce any nature upon a given body, what kind of rule or direction or guidance he would most wish for, and express the same in the simplest and least abstruse language. For instance, if a man wishes to superinduce upon silver that yellow color of gold or an increase of weight (observing the laws of matter), or transparency on an opaque stone, or tenacity on glass, or vegetation on some substance that is not vegetable—we must consider, I say, what kind of rule or guidance he would most desire. And in the first place, he will undoubtedly wish to be directed to something which will not deceive him in the result nor fail him in the trial. Secondly, he will wish for such a rule as shall not tie him down to certain means and particular modes of operation. For perhaps he may not have those means, nor be able conveniently to procure them. And if there be other means and other methods for producing the required nature (besides the one prescribed) these may perhaps be within his reach; and yet he shall be excluded by the narrowness of the rule, and get no good from them. Thirdly, he will desire something to be shown him, which is not as difficult as the thing proposed to be done, but comes nearer to practice.

For a true and perfect rule of operation, then, the direction will be *that it be certain, free, and disposing or leading to action.* And this is the same thing with the discovery of the true form. For the form of a nature is such, that given the form, the nature infallibly follows. Therefore it is always present when the nature is present, and universally implies it, and is constantly inherent in it. Again, the form is such that if it be taken away the nature infallibly vanishes. Therefore it is always absent when the nature is absent, and implies its absence, and inheres in nothing else. Lastly, the true form is such that it deduces the given nature from some source of being which is inherent in more natures, and which is better known in the natural order of things than the form itself. For a true and perfect axiom of knowledge, then, the direction and precept will be, *that another nature be discovered which is convertible with the given nature and yet is a limitation of a more general nature, as of a true and real genus.* Now these two directions, the one active the other contemplative, are one and the same thing; and what in operation is most useful, that in knowledge is most true.

V

The rule or axiom for the transformation of bodies is of two kinds. The first regards a body as a troop or collection of simple natures. In gold, for example, the following properties meet. It is yellow in color, heavy up to a certain weight, malleable or ductile to a certain degree of extension; it is not volatile and loses none of its substance by the action of fire; it turns into a liquid with a certain degree of fluidity; it is separated and dissolved by particular means; and so on for the other natures which meet in gold. This kind of axiom, therefore, deduces the thing from the forms of simple natures. For he who knows the forms of yellow, weight, ductility, fixity, fluidity, solution, and so on, and the methods for superinducing them and their gradations and modes, will make it his care to have them joined together in some body, whence may follow the transformation of that body into gold. And this kind of operation pertains to the first kind of action. For the principle of generating some one simple nature is the same as that of generating many; only that a man is more fettered and tied down in operation, if more are required, by reason of the difficulty of combining into one so many natures which do not readily meet, except in the beaten and ordinary paths of nature. It must be said, however, that this mode of operation (which looks to simple natures though in a compound body) proceeds from what in nature is constant and eternal and universal, and opens broad roads to human power, such as (in the

present state of things) human thought can scarcely comprehend or anticipate.

The second kind of axiom, which is concerned with the
discovery of the *latent process*, proceeds not by simple natures,
but by compound bodies, as they are found in nature in its
ordinary course. As, for instance, when inquiry is made from
what beginnings, and by what method and by what process,
gold or any other metal or stone is generated, from its first
menstrua and rudiments up to the perfect mineral; or in like
manner, by what process herbs are generated, from the first
concretion of juices in the ground or from seeds up to the
formed plant, with all the successive motions and diverse and
continued efforts of nature. So also in the inquiry concerning
the process of development in the generation of animals, from
coition to birth; and in like manner of other bodies.

It is not however only to the generations of bodies that this
investigation extends, but also to other motions and operations of nature. As, for instance, when inquiry is made concerning the whole course and continued action of nutrition,
from the first reception of the food to its complete assimilation; or again, concerning the voluntary motion of animals
from the first impression on the imagination and the continued efforts of the spirit up to the bendings and movements
of the limbs; or concerning the motion of the tongue and lips
and other instruments, and the changes through which it
passes till it comes to the utterance of articulate sounds. For
these inquiries also relate to natures concrete or combined into
one structure, and have regard to what may be called particular and special habits of nature, not to her fundamental and
universal laws which constitute forms. And yet it must be con-
fessed that this plan appears to be readier and to lie nearer at
hand and to give more ground for hope than the primary one.

In like manner the operative which answers to this speculative part, starting from the ordinary incidents of nature, extends its operation to things immediately adjoining, or at least
not far removed. But as for any profound and radical opera-

tions on nature, they depend entirely on the primary axioms.
And in those things too where man has no means of operating,
but only of knowing, as in the heavenly bodies (for these he
cannot operate upon or change or transform), the investigation of the fact itself or truth of the thing, no less than the
knowledge of the causes and consents, must come from those
primary and catholic axioms concerning simple natures, such
as the nature of spontaneous rotation, of attraction or magnetism, and of many others which are of a more general form
than the heavenly bodies themselves. For let no one hope to
decide the question whether it is the earth or heaven that
really revolves in the diurnal motion until he has first comprehended the nature of spontaneous rotation.

VI

But this latent process of which I speak is quite another
thing than men, preoccupied as their minds now are, will
easily conceive. For what I understand by it is not certain
measures or signs or successive steps of process in bodies, which
can be seen; but a process perfectly continuous, which for the
most part escapes the sense.

For instance: in all generation and transformation of bodies,
we must inquire what is lost and escapes; what remains, what
is added; what is expanded, what contracted; what is united,
what separated; what is continued, what cut off; what pro-
pels, what hinders; what predominates, what yields; and a variety of other particulars.

Again, not only in the generation or transformation of
bodies are these points to be ascertained, but also in all other
alterations and motions it should in like manner be inquired
what goes before, what comes after; what is quicker, what
more tardy; what produces, what governs motion; and like
points; all which nevertheless in the present state of the sciences (the texture of which is as rude as possible and good for
nothing) are unknown and unhandled. For seeing that every
natural action depends on things infinitely small, or at least

too small to strike the sense, no one can hope to govern or change nature until he has duly comprehended and observed them.

VII

In like manner the investigation and discovery of the *latent configuration* in bodies is a new thing, no less than the discovery of the latent process and of the form. For as yet we are but lingering in the outer courts of nature, nor are we preparing ourselves a way into her inner chambers. Yet no one can endow a given body with a new nature, or successfully and aptly transmute it into a new body, unless he has attained a competent knowledge of the body so to be altered or transformed. Otherwise he will run into methods which, if not useless, are at any rate difficult and perverse and unsuitable to the nature of the body on which he is operating. It is clear therefore that to this also a way must be opened and laid out.

And it is true that upon the anatomy of organized bodies (as of man and animals) some pains have been well bestowed and with good effect; and a subtle thing it seems to be, and a good scrutiny of nature. Yet this kind of anatomy is subject to sight and sense, and has place only in organized bodies. And besides it is a thing obvious and easy, when compared with the true anatomy of the latent configuration in bodies which are thought to be of uniform structure, especially in things and their parts that have a specific character, as iron, stone; and again in parts of uniform structure in plants and animals, as the root, the leaf, the flower, flesh, blood, and bones. But even in this kind, human industry has not been altogether wanting; for this is the very thing aimed at in the separation of bodies of uniform structure by means of distillations and other modes of analysis; that the complex structure of the compound may be made apparent by bringing together its several homogeneous parts. And this is of use too, and conduces to the object we are seeking, although too often fallacious in its results, because many natures which are in fact newly brought out and superinduced by fire and heat and other modes of solution are taken to be the effect of separation merely, and to have subsisted in the compound before. And after all, this is but a small part of the work of discovering the true configuration in the compound body; which configuration is a thing far more subtle and exact, and such as the operation of fire rather confounds than brings out and makes distinct.

Therefore a separation and solution of bodies must be effected, not by fire indeed, but by reasoning and true induction, with experiments to aid; and by a comparison with other bodies, and a reduction to simple natures and their forms, which meet and mix in the compound. In a word, we must pass from Vulcan to Minerva if we intend to bring to light the true textures and configurations of bodies on which all the occult and, as they are called, specific properties and virtues in things depend, and from which, too, the rule of every powerful alteration and transformation is derived.

For example, we must inquire what amount of spirit there is in every body, what of tangible essence; and of the spirit, whether it be copious and turgid, or meager and scarce; whether it be fine or coarse, akin to air or to fire, brisk or sluggish, weak or strong, progressive or retrograde, interrupted or continuous, agreeing with external and surrounding objects or disagreeing, etc. In like manner we must inquire into the tangible essence (which admits of no fewer differences than the spirit), into its coats, its fibers, its kinds of texture. Moreover, the disposition of the spirit throughout the corporeal frame, with its pores, passages, veins and cells, and the rudiments or first essays of the organized body, falls under the same investigation. But on these inquiries also, and I may say on all the discovery of the latent configuration, a true and clear light is shed by the primary axioms which entirely dispels darkness and subtlety.

VIII

Nor shall we thus be led to the doctrine of atoms, which implies the hypothesis of a vacuum and that of the unchangeableness of matter (both false assumptions); we shall be led only to real particles, such as really exist. Nor again is there any reason to be alarmed at the subtlety of the investigation, as if it could not be disentangled. On the contrary, the nearer it approaches to simple natures, the easier and plainer will everything become, the business being transferred from the complicated to the simple; from the incommensurable to the commensurable; from surds to rational quantities; from the infinite and vague to the finite and certain; as in the case of the letters of the alphabet and the notes of music. And inquiries into nature have the best result when they begin with physics and end in mathematics. Again, let no one be afraid of high numbers or minute fractions. For in dealing with numbers it is as easy to set down or conceive a thousand as one, or the thousandth part of an integer as an integer itself.

IX

From the two kinds of axioms which have been spoken of arises a just division of philosophy and the sciences, taking the received terms (which come nearest to express the thing) in a sense agreeable to my own views. Thus, let the investigation of forms, which are (in the eye of reason at least, and in their essential law) eternal and immutable, constitute *Metaphysics;* and let the investigation of the efficient cause, and of matter, and of the latent process, and the latent configuration (all of which have reference to the common and ordinary course of nature, not to her eternal and fundamental laws) constitute *Physics.* And to these let there be subordinate two practical divisions: to Physics, *Mechanics;* to Metaphysics, what (in a purer sense of the word) I call *Magic,* on account of the broadness of the ways it moves in, and its greater command over nature.

X

Having thus set up the mark of knowledge, we must go on to precepts, and that in the most direct and obvious order. Now my directions for the interpretation of nature embrace two generic divisions: the one how to educe and form axioms from experience; the other how to deduce and derive new experiments from axioms. The former again is divided into three ministrations: a ministration to the sense, a ministration to the memory, and a ministration to the mind or reason.

For first of all we must prepare a natural and experimental history, sufficient and good; and this is the foundation of all, for we are not to imagine or suppose, but to discover, what nature does or may be made to do.

But natural and experimental history is so various and diffuse that it confounds and distracts the understanding, unless it be ranged and presented to view in a suitable order. We must therefore form tables and arrangements of instances, in such a method and order that the understanding may be able to deal with them.

And even when this is done, still the understanding, if left to itself and its own spontaneous movements, is incompetent and unfit to form axioms, unless it be directed and guarded. Therefore in the third place we must use induction, true and legitimate induction, which is the very key of interpretation. But of this, which is the last, I must speak first, and then go back to the other ministrations.

XI

The investigation of forms proceeds thus: a nature being given, we must first of all have a muster or presentation before the understanding of all known instances which agree in the same nature, though in substances the most unlike. And such collection must be made in the manner of a history, without premature speculation, or any great amount of subtlety. For example, let the investigation be into the form of heat.

Instances Agreeing in the Nature of Heat

1. The rays of the sun, especially in summer and at noon.

2. The rays of the sun reflected and condensed, as between mountains, or on walls, and most of all in burning glasses and mirrors.

3. Fiery meteors.

4. Burning thunderbolts.

5. Eruptions of flame from the cavities of mountains.

6. All flame.

7. Ignited solids.

8. Natural warm baths.

9. Liquids boiling or heated.

10. Hot vapors and fumes, and the air itself, which conceives the most powerful and glowing heat if confined, as in reverbatory furnaces.

11. Certain seasons that are fine and cloudless by the constitution of the air itself, without regard to the time of year.

12. Air confined and underground in some caverns, especially in winter.

13. All villous substances, as wool, skins of animals, and down of birds, have heat.

14. All bodies, whether solid or liquid, whether dense or rare (as the air itself is), held for a time near the fire.

15. Sparks struck from flint and steel by strong percussion.

16. All bodies rubbed violently, as stone, wood, cloth, etc., insomuch that poles and axles of wheels sometimes catch fire; and the way they kindled fire in the West Indies was by attrition.

17. Green and moist vegetables confined and bruised together, as roses packed in baskets; insomuch that hay, if damp, when stacked, often catches fire.

18. Quicklime sprinkled with water.

19. Iron, when first dissolved by strong waters in glass, and that without being put near the fire. And in like manner tin, etc., but not with equal intensity.

20. Animals, especially and at all times internally; though

in insects the heat is not perceptible to the touch by reason of the smallness of their size.

21. Horse dung and like excrements of animals, when fresh.

22. Strong oil of sulphur and of vitriol has the effect of heat in burning linen.

23. Oil of marjoram and similar oils have the effect of heat in burning the bones of the teeth.

24. Strong and well rectified spirit of wine has the effect of heat, insomuch that the white of an egg being put into it hardens and whitens almost as if it were boiled, and bread thrown in becomes dry and crusted like toast.

25. Aromatic and hot herbs, as *dracunculus, nasturtium vetus,* etc., although not warm to the hand (either whole or in powder), yet to the tongue and palate, being a little masticated, they feel hot and burning.

26. Strong vinegar, and all acids, on all parts of the body where there is no epidermis, as the eye, tongue, or on any part when wounded and laid bare of the skin, produce a pain but little differing from that which is created by heat.

27. Even keen and intense cold produces a kind of sensation of burning: "Nec Boreæ penetrabile frigus adurit." [1]

28. Other instances.

This table I call the *Table of Essence and Presence.*

XII

Secondly, we must make a presentation to the understanding of instances in which the given nature is wanting; because the form, as stated above, ought no less to be absent when the given nature is absent, than present when it is present. But to note all these would be endless.

The negatives should therefore be subjoined to the affirmatives, and the absence of the given nature inquired of in those subjects only that are most akin to the others in which it is present and forthcoming. This I call the *Table of Deviation, or of Absence in Proximity.*

[1] Nor burns the sharp cold of the northern blast.

Instances in Proximity where the Nature of Heat is Absent

Answering to the first affirmative instance.

1. The rays of the moon and of stars and comets are not found to be hot to the touch; indeed the severest colds are observed to be at the full moons.

The larger fixed stars, however, when passed or approached by the sun, are supposed to increase and give intensity to the heat of the sun, as is the case when the sun is in the sign Leo, and in the dog days.

To the 2nd.

2. The rays of the sun in what is called the middle region of the air do not give heat; for which there is commonly assigned not a bad reason, viz., that that region is neither near enough to the body of the sun from which the rays emanate, nor to the earth from which they are reflected. And this appears from the fact that on the tops of mountains, unless they are very high, there is perpetual snow. On the other hand, it has been observed that on the Peak of Tenerife, and among the Andes of Peru, the very tops of the mountains are free from snow, which lies only somewhat lower down. Moreover, the air itself at the very top is found to be by no means cold, but only rare and keen; insomuch that on the Andes it pricks and hurts the eyes by its excessive keenness, and also irritates the mouth of the stomach, producing vomiting. And it was observed by the ancients that on the top of Olympus the rarity of the air was such that those who ascended it had to carry sponges with them dipped in vinegar and water, and to apply them from time to time to the mouth and nose, the air being from its rarity not sufficient to support respiration; and it was further stated that on this summit the air was so serene, and so free from rain and snow and wind, that letters traced by the finger in the ashes of the sacrifices on the altar of Jupiter remained there still the next year without being at all disturbed. And at this day travelers ascending to the top of the Peak of Tenerife make the ascent by night and not by day, and soon after the rising of the sun

are warned and urged by their guides to come down without delay, on account of the danger they run lest the animal spirits should swoon and be suffocated by the tenuity of the air.

To the 2nd. 3. The reflection of the rays of the sun in regions near the polar circles is found to be very weak and ineffective in producing heat, insomuch that the Dutch who wintered in Nova Zembla and expected their ship to be freed from the obstructions of the mass of ice which hemmed her in by the beginning of July, were disappointed in their expectation and obliged to take to their boat. Thus the direct rays of the sun seem to have but little power, even on the level ground; nor have the reflex much, unless they are multiplied and combined, which is the case when the sun tends more to the perpendicular, for then the incident rays make acuter angles, so that the lines of the rays are nearer each other; whereas on the contrary, when the sun shines very obliquely, the angles are very obtuse, and thus the lines of rays are at a greater distance from each other. Meanwhile, it should be observed that there may be many operations of the sun, and those too depending on the nature of heat, which are not proportioned to our touch, so that in respect to us their action does not go so far as to produce sensible warmth, but in respect to some other bodies they have the effect of heat.

To the 2nd. 4. Try the following experiment. Take a glass fashioned in a contrary manner to a common burning glass and, placing it between your hand and the rays of the sun, observe whether it diminishes the heat of the sun, as a burning glass increases and strengthens it. For it is evident in the case of optical rays that according as the glass is made thicker or thinner in the middle as compared with the sides, so do the objects seen through it appear more spread or more contracted. Observe therefore whether the same is the case with heat.

To the 2nd. 5. Let the experiment be carefully tried, whether by means of the most powerful and best constructed burning glasses, the rays of the moon can be so caught and collected as to produce even the last degree of warmth. But should this degree of warmth prove too subtle and weak to be perceived and apprehended by the touch, recourse must be had to those glasses which indicate the state of the atmosphere in respect to heat and cold. Thus, let the rays of the moon fall through a burning glass on the top of a glass of this kind, and then observe whether there ensues a sinking of the water through warmth.

To the 2nd.   6. Let a burning glass also be tried with a heat that does not emit rays or light, as that of iron or stone heated but not ignited, boiling water, and the like; and observe whether there ensue an increase of the heat, as in the case of the sun's rays.

To the 2nd.   7. Let a burning glass also be tried with common flame.

To the 3rd.   8. Comets (if we are to reckon these too among meteors) are not found to exert a constant or manifest effect in increasing the heat of the season, though it is observed that they are often followed by droughts. Moreover bright beams and pillars and openings in the heavens appear more frequently in winter than in summertime, and chiefly during the intensest cold, but always accompanied by dry weather. Lightning, however, and coruscations and thunder seldom occur in the winter, but about the time of great heat. Falling stars, as they are called, are commonly supposed to consist rather of some bright and lighted viscous substance, than to be of any strong fiery nature. But on this point let further inquiry be made.

To the 4th.   9. There are certain coruscations which give light but do not burn. And these always come without thunder.

To the 5th.   10. Eructations and eruptions of flame are found no less in cold than in warm countries, as in Iceland and Greenland. In cold countries, too, the trees are in many cases more inflammable and more pitchy and resinous than in warm; as the fir, pine, and others. The situations however and the nature of the soil in which eruptions

of this kind usually occur have not been carefully enough ascertained to enable us to subjoin a negative to this affirmative instance.

To the 6th.   11. All flame is in all cases more or less warm;
nor is there any negative to be subjoined.
And yet they say that the *ignis fatuus* (as it is called), which
sometimes even settles on a wall, has not much heat, perhaps
as much as the flame of spirit of wine, which is mild and soft.
But still milder must that flame be which, according to certain grave and trustworthy histories has been seen shining
about the head and locks of boys and girls, without at all
burning the hair, but softly playing round it. It is also most
certain that about a horse, when sweating on the road, there
is sometimes seen at night, and in clear weather, a sort of
luminous appearance without any manifest heat. And it is a
well-known fact, and looked upon as a sort of miracle, that a
few years ago a girl's stomacher, on being slightly shaken or
rubbed, emitted sparks, which was caused perhaps by some
alum or salts used in the dye, that stood somewhat thick and
formed a crust, and were broken by the friction. It is also
most certain that all sugar, whether refined or raw, provided
only it be somewhat hard, sparkles when broken or scraped
with a knife in the dark. In like manner sea and salt water is
sometimes found to sparkle by night when struck violently by
oars. And in storms, too, at nighttime, the foam of the sea
when violently agitated emits sparks, and this sparkling the
Spaniards call *Sea Lung*. With regard to the heat of the flame
which was called by ancient sailors Castor and Pollux, and by
moderns St. Elmo's Fire, no sufficient investigation thereof has
been made.

To the 7th.   12. Every body ignited so as to turn to a fiery
red, even if unaccompanied by flame, is al-
ways hot; neither is there any negative to be subjoined to this
affirmative. But that which comes nearest seems to be rotten
wood, which shines by night and yet is not found to be hot;
and the putrefying scales of fish, which also shine in the dark
and yet are not warm to the touch; nor, again, is the body of

the glowworm, or of the fly called *Luciola*, found to be warm to the touch.

To the 8th.   13. In what situation and kind of soil warm baths usually spring has not been sufficiently examined; and therefore no negative is subjoined.

To the 9th.   14. To warm liquids I subjoin the negative instance of liquid itself in its natural state. For we find no tangible liquid which is warm in its own nature and remains so constantly; but the warmth is of an adventitious nature, superinduced only for the time being, so that the liquids which in power and operation are hottest, as spirit of wine, chemical oil of spices, oil of vitriol and sulphur, and the like, which burn after a while, are at first cold to the touch. The water of natural warm baths, on the other hand, if received into a vessel and separated from its springs, cools just like water that has been heated on a fire. But it is true that oily substances are less cold to the touch than watery, oil being less cold than water, and silk than linen. But this belongs to the Table of Degrees of Cold.

To the 10th.   15. In like manner to hot vapor I subjoin as a negative the nature of vapor itself, such as we find it with us. For exhalations from oily substances, though easily inflammable, are yet not found to be warm unless newly exhaled from the warm body.

To the 10th.   16. In like manner I subjoin as a negative to hot air the nature of air itself. For we do not find here any air that is warm, unless it has either been confined, or compressed, or manifestly warmed by the sun, fire, or some other warm substance.

To tne 11th.   17. I here subjoin the negative of colder weather than is suitable to the season of the year, which we find occurs during east and north winds; just as we have weather of the opposite kind with the south and west winds. So a tendency to rain, especially in wintertime, accompanies warm weather; while frost accompanies cold.

To the 12th.   18. Here I subjoin the negative of air confined in caverns during the summer. But the subject of air in confinement should by all means be more diligently examined. For in the first place it may well be a matter of doubt what is the nature of air in itself with regard to heat and cold. For air manifestly receives warmth from the influence of the heavenly bodies, and cold perhaps from the exhalations of the earth; and, again, in the middle region of air, as it is called, from cold vapors and snow. So that no opinion can be formed as to the nature of air from the examination of air that is at large and exposed, but a truer judgment might be made by examining it when confined. It is, however, necessary for the air to be confined in a vessel of such material as will not itself communicate warmth or cold to the air by its own nature, nor readily admit the influence of the outer atmosphere. Let the experiment therefore be made in an earthen jar wrapped round with many folds of leather to protect it from the outward air, and let the vessel remain tightly closed for three or four days; then open the vessel and test the degree of heat or cold by applying either the hand or a graduated glass.

To the 13th.   19. In like manner a doubt suggests itself whether the warmth in wool, skins, feathers, and the like, proceeds from a faint degree of heat inherent in them, as being excretions from animals; or from a certain fat and oiliness, which is of a nature akin to warmth; or simply, as surmised in the preceding article, from the confinement and separation of the air. For all air that is cut off from connection with the outer air seems to have some warmth. Try the experiment therefore with fibrous substances made of linen; not of wool, feathers, or silk, which are excretions from animals. It should also be observed that all powders (in which there is manifestly air enclosed) are less cold than the whole substances they are made from; as likewise I suppose that all froth (as that which contains air) is less cold than the liquor it comes from.

To the 14th.   20. To this no negative is subjoined. For there is nothing found among us, either tangible or spirituous, which does not contract warmth when put near fire. There is this difference however, that some substances contract warmth more quickly, as air, oil, and water; others more slowly, as stone and metal. But this belongs to the Table of Degrees.

To the 15th.
21. To this instance I subjoin no negative, except that I would have it well observed that sparks are produced from flint and steel, or any other hard substance, only when certain minute particles are struck off from the substance of the stone or metal; and that the attrition of the air does not of itself ever produce sparks, as is commonly supposed. And the sparks themselves, too, owing to the weight of the ignited body, tend rather downwards than upwards; and on going out become a tangible sooty substance.

To the 16th.
22. There is no negative, I think, to be subjoined to this instance. For we find among us no tangible body which does not manifestly gain warmth by attrition; insomuch that the ancients fancied that the heavenly bodies had no other means or power of producing warmth than by the attrition of the air in their rapid and hurried revolution. But on this subject we must further inquire whether bodies discharged from engines, as balls from cannon, do not acquire some degree of heat from the very percussion, so as to be found somewhat warm when they fall. Air in motion, however, rather chills than warms, as appears from wind, bellows, and blowing with the mouth contracted. But motion of this kind is not so rapid as to excite heat, and is the motion of a mass, and not of particles; so that it is no wonder if it does not generate heat.

To the 17th.
23. On this instance should be made more diligent inquiry. For herbs and vegetables when green and moist seem to contain some latent heat, though so slight that it is not perceptible to the touch when they are single, but only when they are collected and shut up together, so that their spirits may not breathe out into the air, but may mutually cherish each other; whereupon there arises a palpable heat, and sometimes flame in suitable matter.

24. On this instance too should be made more diligent inquiry. For quicklime sprinkled with water seems to contract heat either by the concentration of heat before dispersed, as in the above-mentioned case of confined herbs, or because the igneous spirit is irritated and exasperated by the water so as to cause a conflict and reaction. Which of these two is the real cause will more readily appear if oil be poured on instead of water, for oil will serve equally well with water to concentrate the enclosed spirit, but not to irritate it. We should also extend the experiment both by employing the ashes and rusts of different bodies, and by pouring in different liquids.

To the 18th.

25. To this instance is subjoined the negative of other metals which are softer and more fusible. For gold leaf dissolved by *aqua regia* gives no heat to the touch; no more does lead dissolved in *aqua fortis;* neither again does quicksilver (as I remember); but silver itself does, and copper too (as I remember); tin still more manifestly; and most of all iron and steel, which not only excite a strong heat in dissolution but also a violent ebullition. It appears therefore that the heat is produced by conflict, the strong waters penetrating, digging into, and tearing asunder the parts of the substance, while the substance itself resists. But where the substances yield more easily, there is hardly any heat excited.

To the 19th.

26. To the heat of animals no negative is subjoined, except that of insects (as above-mentioned) on account of their small size. For in fishes, as compared with land animals, it is rather a low degree than an absence of heat that is noted. But in vegetables and plants there is no degree of heat perceptible to the touch, either in their exudations or in their pith when freshly exposed. In animals, however, is found a great diversity of heat, both in their parts (there being different degrees of heat about the heart, in the brain, and on the skin) and in their accidents, as violent exercise and fevers.

To the 20th.

To the 21st. 27. To this instance it is hard to subjoin a negative. Indeed the excrements of animals when no longer fresh have manifestly a potential heat, as is seen in the enriching of soil.

To the 22nd and 23rd. 28. Liquids, whether waters or oils, which possess a great and intense acridity, act like heat in tearing asunder bodies and burning them after some time; yet to the touch they are not hot at first. But their operation is relative and according to the porosity of the body to which they are applied. For *aqua regia* dissolves gold but not silver; *aqua fortis,* on the contrary, dissolves silver, but not gold; neither dissolves glass, and so on with others.

To the 24th. 29. Let trial be made of spirit of wine on wood, and also on butter, wax, or pitch; and observe whether by its heat it in any degree melts them. For the twenty-fourth instance exhibits a power in it that resembles heat in producing incrustation. In like manner therefore try its power in producing liquefaction. Let trial also be made with a graduated or calendar glass, hollow at the top; pour into the hollow spirit of wine well rectified, cover it up that the spirit may better retain its heat, and observe whether by its heat it makes the water sink.

To the 25th. 30. Spices and acrid herbs strike hot on the palate, and much hotter on the stomach. Observe therefore on what other substances they produce the effects of heat. Sailors tell us that when large parcels and masses of spices are, after being long kept close, suddenly opened, those who first stir and take them out run the risk of fever and inflammation. It can also be tried whether such spices and herbs when pounded would not dry bacon and meat hung over them, as smoke does.

To the 26th. 31. There is an acridity or pungency both in cold things, as vinegar and oil of vitriol, and in hot, as oil of marjoram and the like. Both alike therefore cause pain in animate substances, and tear asunder and consume the parts in such as are inanimate. To this instance again there is no negative subjoined. Moreover we find no pain in animals, save with a certain sensation of heat.

To the 27th.

32. There are many actions common both to heat and cold, though in a very different manner. For boys find that snow after a while seems to burn their hands; and cold preserves meat from putrefaction, no less than fire; and heat contracts bodies, which cold does also. But these and similar instances may more conveniently be referred to the inquiry concerning cold.

XIII

Thirdly, we must make a presentation to the understanding of instances in which the nature under inquiry is found in different degrees, more or less; which must be done by making a comparison either of its increase and decrease in the same subject, or of its amount in different subjects, as compared one with another. For since the form of a thing is the very thing itself, and the thing differs from the form no otherwise than as the apparent differs from the real, or the external from the internal, or the thing in reference to man from the thing in reference to the universe, it necessarily follows that no nature can be taken as the true form, unless it always decrease when the nature in question decreases, and in like manner always increase when the nature in question increases. This Table therefore I call the *Table of Degrees* or the *Table of Comparison.*

Table of Degrees or Comparison in Heat

I will therefore first speak of those substances which contain no degree at all of heat perceptible to the touch, but seem to have a certain potential heat only, or disposition and preparation for hotness. After that I shall proceed to substances which are hot actually, and to the touch, and to their intensities and degrees.

1. In solid and tangible bodies we find nothing which is in its nature originally hot. For no stone, metal, sulphur, fossil, wood, water, or carcass of animal is found to be hot. And the hot water in baths seems to be heated by external causes; whether it be by flame or subterraneous fire, such as is thrown up from Etna and many other mountains, or by the conflict of bodies, as heat is caused in the dissolution of iron and tin. There is therefore no degree of heat palpable to the touch in animate substances; but they differ in degree of cold, wood not being equally cold with metal. But this belongs to the Table of Degrees in Cold.

2. As far, however, as potential heat and aptitude for flame is concerned, there are many inanimate substances found strongly disposed thereto, as sulphur, naphtha, rock oil.

3. Substances once hot, as horse dung from animal heat, and lime or perhaps ashes and soot from fire, retain some latent remains of their former heat. Hence certain distillations and resolutions of bodies are made by burying them in horse dung, and heat is excited in lime by sprinkling it with water, as already mentioned.

4. In the vegetable creation we find no plant or part of plant (as gum or pitch) which is warm to the human touch. But yet, as stated above, green herbs gain warmth by being shut up; and to the internal touch, as the palate or stomach, and even to external parts, after a little time, as in plasters and ointments, some vegetables are perceptibly warm and others cold.

5. In the parts of animals after death or separation from the body, we find nothing warm to the human touch. Not even horse dung, unless enclosed and buried, retains its heat. But yet all dung seems to have a potential heat, as is seen in the fattening of the land. In like manner carcasses of animals have some such latent and potential heat, insomuch that in burying grounds, where burials take place daily, the earth collects a certain hidden heat which consumes a body newly laid in it much more speedily than pure earth. We are told too that in the East there is discovered a fine soft texture,

made of the down of birds, which by an innate force dissolves and melts butter when lightly wrapped in it.

6. Substances which fatten the soil, as dung of all kinds, chalk, sea sand, salt, and the like, have some disposition to heat.

7. All putrefaction contains in itself certain elements of a slight heat, though not so much as to be perceived by the touch. For not even those substances which on putrefaction turn to animalculae, as flesh, cheese, etc., feel warm to the touch; no more does rotten wood, which shines in the dark. Heat, however, in putrid substances sometimes betrays itself by foul and powerful odors.

8. The first degree of heat therefore among those substances which feel hot to the touch, seems to be the heat of animals, which has a pretty great extent in its degrees. For the lowest, as in insects, is hardly perceptible to the touch, but the highest scarcely equals the sun's heat in the hottest countries and seasons, nor is it too great to be borne by the hand. It is said, however, of Constantius, and some others of a very dry constitution and habit of body, that in violent fevers they became so hot as somewhat to burn the hand that touched them.

9. Animals increase in heat by motion and exercise, wine, feasting, venus, burning fevers, and pain.

10. When attacked by intermittent fevers, animals are at first seized with cold and shivering, but soon after they become exceedingly hot, which is their condition from the first in burning and pestilential fevers.

11. Let further inquiry be made into the different degrees of heat in different animals, as in fishes, quadrupeds, serpents, birds; and also according to their species, as in the lion, the kite, the man; for in common opinion fish are the least hot internally, and birds the hottest, especially doves, hawks, and sparrows.

12. Let further inquiry be made into the different degrees of heat in the different parts and limbs of the same animal. For milk, blood, seed, eggs, are found to be hot only in a moderate degree, and less hot than the outer flesh of the ani-

mal when in motion or agitated. But what the degree of heat
is in the brain, stomach, heart, etc., has not yet been in like
manner inquired.

13. All animals in winter and cold weather are cold externally, but internally they are thought to be even hotter.

14. The heat of the heavenly bodies, even in the hottest
countries, and at the hottest times of the year and day, is
never sufficiently strong to set on fire or burn the driest wood
or straw, or even tinder, unless strengthened by burning
glasses or mirrors. It is, however, able to extract vapor from
moist substances.

15. By the tradition of astronomers some stars are hotter
than others. Of planets, Mars is accounted the hottest after
the sun; then comes Jupiter, and then Venus. Others, again,
are set down as cold: the moon, for instance, and above all
Saturn. Of fixed stars, Sirius is said to be the hottest, then Cor
Leonis or Regulus, then Canicula, and so on.

16. The sun gives greater heat the nearer he approaches to
the perpendicular or zenith; and this is probably true of the
other planets also, according to the proportion of their heat.
Jupiter, for instance, is hotter, probably, to us when under
Cancer or Leo than under Capricorn or Aquarius.

17. We must also believe that the sun and other planets
give more heat in perigee, from their proximity to the earth,
than they do in apogee. But if it happens that in some region
the sun is at the same time in perigee and near the perpendicular, his heat must of necessity be greater than in a region
where he is also in perigee, but shining more obliquely. And
therefore the altitude of the planets in their exaltation in different regions ought to be noted, with respect to perpendicularity or obliquity.

18. The sun and other planets are supposed to give greater
heat when nearer to the larger fixed stars. Thus when the sun
is in Leo he is nearer Cor Leonis, Cauda Leonis, Spica Vir-
ginis, Sirius and Canicula, than when he is in Cancer, in
which sign, however, he is nearer to the perpendicular. And it
must be supposed that those parts of the heavens shed the

greatest heat (though it be not at all perceptible to the touch) which are the most adorned with stars, especially of a larger size.

19. Altogether, the heat of the heavenly bodies is increased in three ways: first, by perpendicularity; secondly, by proximity or perigee; thirdly, by the conjunction or combination of stars.

20. The heat of animals, and of the rays of the heavenly bodies also (as they reach us), is found to differ by a wide interval from flame, though of the mildest kind, and from all ignited bodies; and from liquids also, and air itself when highly heated by fire. For the flame of spirit of wine, though scattered and not condensed, is yet sufficient to set paper, straw, or linen on fire, which the heat of animals will never do, or of the sun without a burning glass or mirror.

21. There are, however, many degrees of strength and weakness in the heat of flame and ignited bodies. But as they have never been diligently inquired into, we must pass them lightly over. It appears, however, that of all flame that of spirit of wine is the softest, unless perhaps *ignis fatuus* be softer, and the flames or sparklings arising from the sweat of animals. Next to this, as I suppose, comes flame from light and porous vegetable matter, as straw, reeds, and dried leaves, from which the flame from hairs or feathers does not much differ. Next perhaps comes flame from wood, especially such as contains but little rosin or pitch; with this distinction, however, that the flame from small pieces of wood (such as are commonly tied up in fagots) is milder than the flame from trunks and roots of trees. And this you may try any day in furnaces for smelting iron, in which a fire made with fagots and boughs of trees is of no great use. After this I think comes flame from oil, tallow, wax, and such like fat and oily substances, which have no great acrimony. But the most violent heat is found in pitch and rosin; and yet more in sulphur, camphor, naphtha, rock oil, and salts (after the crude matter is discharged), and in their compounds, as gunpowder, Greek fire (commonly called

wildfire), and its different kinds, which have so stubborn a heat that they are not easily extinguished by water.

22. I think also that the flame which results from some imperfect metals is very strong and eager. But on these points let further inquiry be made.

23. The flame of powerful lightning seems to exceed in strength all the former, for it has even been known to melt wrought iron into drops, which those other flames cannot do.

24. In ignited bodies too there are different degrees of heat, though these again have not yet been diligently examined. The weakest heat of all, I think, is that from tinder, such as we use to kindle flame with; and in like manner that of touchwood or tow, which is used in firing cannon. After this comes ignited wood or coal, and also bricks and the like heated to ignition. But of all ignited substances, the hottest, as I take it, are ignited metals, as iron, copper, etc. But these require further investigation.

25. Some ignited bodies are found to be much hotter than some flames. Ignited iron, for instance, is much hotter and more consuming than flame of spirit of wine.

26. Of substances also which are not ignited but only heated by fire, as boiling water and air confined in furnaces, some are found to exceed in heat many flames and ignited substances.

27. Motion increases heat, as you may see in bellows and by blowing; insomuch that the harder metals are not dissolved or melted by a dead or quiet fire, till it be made intense by blowing.

28. Let trial be made with burning glasses, which (as I remember) act thus. If you place a burning glass at the distance of (say) a span from a combustible body, it will not burn or consume it so easily as if it were first placed at the distance of (say) half a span, and then moved gradually and slowly to the distance of the whole span. And yet the cone and union of rays are the same; but the motion itself increases the operation of the heat.

29. Fires which break out during a strong wind are thought to make greater progress against than with it; because the flame recoils more violently when the wind gives way than it advances while the wind is driving it on.

30. Flame does not burst out, nor is it generated, unless some hollow space be allowed it to move and play in; except the explosive flame of gunpowder and the like, where compression and imprisonment increase its fury.

31. An anvil grows very hot under the hammer, insomuch that if it were made of a thin plate it might, I suppose, with strong and continuous blows of the hammer, grow red like ignited iron. But let this be tried by experiment.

32. But in ignited substances which are porous, so as to give the fire room to move, if this motion be checked by strong compression, the fire is immediately extinguished. For instance, when tinder, or the burning wick of a candle or lamp, or even live charcoal or coal, is pressed down with an extinguisher, or with the foot, or any similar instrument, the operation of the fire instantly ceases.

33. Approximation to a hot body increases heat in proportion to the degree of approximation. And this is the case also with light; for the nearer an object is brought to the light, the more visible it becomes.

34. The union of different heats increases heat, unless the hot substances be mixed together. For a large fire and a small fire in the same room increase one another's heat; but warm water plunged into boiling water cools it.

35. The continued application of a hot body increases heat, because heat perpetually passing and emanating from it mingles with the previously existing heat, and so multiplies the heat. For a fire does not warm a room as well in half an hour as it does if continued through the whole hour. But this is not the case with light; for a lamp or candle gives no more light after it has been long lighted than it did at first.

36. Irritation by surrounding cold increases heat, as you may see in fires during a sharp frost. And this I think is owing not merely to the confinement and contraction of the heat,

which is a kind of union, but also to irritation. Thus, when air or a stick is violently compressed or bent, it recoils not merely to the point it was forced from, but beyond it on the other side. Let trial therefore be carefully made by putting a stick or some such thing into flame, and observing whether it is not burnt more quickly at the sides than in the middle of the flame.

37. There are many degrees in susceptibility of heat. And first of all it is to be observed how slight and faint a heat changes and somewhat warms even those bodies which are least of all susceptible of heat. Even the heat of the hand communicates some heat to a ball of lead or any metal, if held in it a little while. So readily and so universally is heat transmitted and excited, the body remaining to all appearance unchanged.

38. Of all substances that we are acquainted with, the one which most readily receives and loses heat is air; as is best seen in calendar glasses [air thermoscopes], which are made thus. Take a glass with a hollow belly, a thin and oblong neck; turn it upside down and lower it, with the mouth downwards and the belly upwards, into another glass vessel containing water; and let the mouth of the inserted vessel touch the bottom of the receiving vessel and its neck lean slightly against the mouth of the other, so that it can stand. And that this may be done more conveniently, apply a little wax to the mouth of the receiving glass, but not so as to seal its mouth quite up, in order that the motion, of which we are going to speak, and which is very facile and delicate, may not be impeded by want of a supply of air.

The lowered glass, before being inserted into the other, must be heated before a fire in its upper part, that is its belly. Now when it is placed in the position I have described, the air which was dilated by the heat will, after a lapse of time sufficient to allow for the extinction of that adventitious heat, withdraw and contract itself to the same extension or dimension as that of the surrounding air at the time of the immersion of the glass, and will draw the water upwards to a corresponding height. To the side of the glass there should be affixed a strip of paper, narrow and oblong, and marked with as many degrees as you choose. You will then see, according as the day is warm or cold, that the air contracts under the action of cold, and expands under the action of heat; as will be seen by the water rising when the air contracts, and sinking when it dilates. But the air's sense of heat and cold is so subtle and exquisite as far to exceed the perception of the human touch, insomuch that a ray of sunshine, or the heat of the breath, much more the heat of one's hand placed on the top of the glass, will cause the water immediately to sink in a perceptible degree. And yet I think that animal spirits have a sense of heat and cold more exquisite still, were it not that it is impeded and deadened by the grossness of the body.

39. Next to air, I take those bodies to be most sensitive to heat which have been recently changed and compressed by cold, as snow and ice; for they begin to dissolve and melt with any gentle heat. Next to them, perhaps, comes quicksilver. After that follow greasy substances, as oil, butter, and the like; then comes wood; then water; and lastly stones and metals, which are slow to heat, especially in the inside. These, however, when once they have acquired heat retain it very long; in so much that an ignited brick, stone, or piece of iron, when plunged into a basin of water, will remain for a quarter of an hour, or thereabouts, so hot that you cannot touch it.

40. The less the mass of a body, the sooner is it heated by the approach of a hot body; which shows that all heat of which we have experience is in some sort opposed to tangible matter.

41. Heat, as far as regards the sense and touch of man, is a thing various and relative; insomuch that tepid water feels hot if the hand be cold, but cold if the hand be hot.

XIV

How poor we are in history anyone may see from the foregoing tables, where I not only insert sometimes mere traditions and reports (though never without a note of doubtful credit and authority) in place of history proved and instances certain, but am also frequently forced to use the words "Let trial be made," or "Let it be further inquired."

XV

The work and office of these three tables I call the Presentation of Instances to the Understanding. Which presentation having been made, induction itself must be set at work; for the problem is, upon a review of the instances, all and each, to find such a nature as is always present or absent with the given nature, and always increases and decreases with it; and which is, as I have said, a particular case of a more general nature. Now if the mind attempt this affirmatively from the first, as when left to itself it is always wont to do, the result will be fancies and guesses and notions ill defined, and axioms that must be mended every day, unless like the schoolmen we have a mind to fight for what is false; though doubtless these will be better or worse according to the faculties and strength of the understanding which is at work. To God, truly, the Giver and Architect of Forms, and it may be to the angels and higher intelligences, it belongs to have an affirmative knowledge of forms immediately, and from the first contemplation. But this assuredly is more than man can do, to whom it is granted only to proceed at first by negatives, and at last to end in affirmatives after exclusion has been exhausted.

XVI

We must make, therefore, a complete solution and separation of nature, not indeed by fire, but by the mind, which is a kind of divine fire. The first work, therefore, of true induction (as far as regards the discovery of forms) is the rejection or exclusion of the several natures which are not found in some instance where the given nature is present, or are found in some instance where the given nature is absent, or are found to increase in some instance when the given nature decreases, or to decrease when the given nature increases. Then indeed after the rejection and exclusion has been duly made, there will remain at the bottom, all light opinions vanishing into smoke, a form affirmative, solid, and true and well defined. This is quickly said; but the way to come at it is winding and intricate. I will endeavor, however, not to overlook any of the points which may help us toward it.

XVII

But when I assign so prominent a part to forms, I cannot too often warn and admonish men against applying what I say to those forms to which their thoughts and contemplations have hitherto been accustomed.

For in the first place I do not at present speak of compound forms, which are, as I have remarked, combinations of simple natures according to the common course of the universe: as of the lion, eagle, rose, gold, and the like. It will be time to treat of these when we come to the latent processes and latent configurations, and the discovery of them, as they are found in what are called substances or natures concrete.

And even in the case of simple natures I would not be understood to speak of abstract forms and ideas, either not defined in matter at all, or ill defined. For when I speak of forms, I mean nothing more than those laws and determinations of absolute actuality which govern and constitute any simple nature, as heat, light, weight, in every kind of matter and subject that is susceptible of them. Thus the form of heat or the form of light is the same thing as the law of heat or the law of light. Nor indeed do I ever allow myself to be drawn away from things themselves and the operative part. And therefore when I say (for instance) in the investigation of the form of heat, "reject rarity," or "rarity does not belong to the form of heat," it is the same as if I said, "It is possible to superinduce heat on a dense body"; or, "It is possible to take away or keep out heat from a rare body."

But if anyone conceive that my forms too are of a somewhat abstract nature, because they mix and combine things heterogeneous (for the heat of heavenly bodies and the heat of fire seem to be very heterogeneous; so do the fixed red of the rose or the like, and the apparent red in the rainbow, the opal, or the diamond; so again do the different kinds of death: death by drowning, by hanging, by stabbing, by apoplexy, by atrophy; and yet they agree severally in the nature of heat, redness, death); if anyone, I say, be of this opinion, he may be assured that his mind is held in captivity by custom, by the gross appearance of things, and by men's opinions. For it is most certain that these things, however heterogeneous and alien from each other, agree in the form or law which governs heat, redness and death; and that the power of man cannot possibly be emancipated and freed from the common course of nature, and expanded and exalted to new efficients and new modes of operation, except by the revelation and discovery of forms of this kind. And yet, when I have spoken of this union of nature, which is the point of most importance, I shall proceed to the divisions and veins of nature, as well the ordinary as those that are more inward and exact, and speak of them in their place.

XVIII

I must now give an example of the exclusion or rejection of natures which by the Tables of Presentation are found not to belong to the form of heat; observing in the meantime that not only each table suffices for the rejection of any nature, but even any one of the particular instances contained in any of the tables. For it is manifest from what has been said that any one contradictory instance overthrows a conjecture as to the form. But nevertheless for clearness' sake and that the use of the tables may be more plainly shown, I sometimes double or multiply an exclusion.

*An Example of Exclusion, or Rejection of Natures
from the Form of Heat*

1. On account of the rays of the sun, reject the nature of the elements.

2. On account of common fire, and chiefly subterraneous fires (which are the most remote and most completely separate from the rays of heavenly bodies), reject the nature of heavenly bodies.

3. On account of the warmth acquired by all kinds of bodies (minerals, vegetables, skin of animals, water, oil, air, and the rest) by mere approach to a fire, or other hot body, reject the distinctive or more subtle texture of bodies.

4. On account of ignited iron and other metals, which communicate heat to other bodies and yet lose none of their weight or substance, reject the communication or admixture of the substance of another hot body.

5. On account of boiling water and air, and also on account of metals and other solids that receive heat but not to ignition or red heat, reject light or brightness.

6. On account of the rays of the moon and other heavenly bodies, with the exception of the sun, also reject light and brightness.

7. By a comparison of ignited iron and the flame of spirit of wine (of which ignited iron has more heat and less brightness, while the flame of spirit of wine has more brightness and less heat), also reject light and brightness.

8. On account of ignited gold and other metals, which are of the greatest density as a whole, reject rarity.

9. On account of air, which is found for the most part cold and yet remains rare, also reject rarity.

10. On account of ignited iron, which does not swell in bulk, but keeps within the same visible dimensions, reject local or expansive motion of the body as a whole.

11. On account of the dilation of air in calendar glasses and the like, wherein the air evidently moves locally and expan-

sively and yet acquires no manifest increase of heat, also reject local or expansive motion of the body as a whole.

12. On account of the ease with which all bodies are heated, without any destruction or observable alteration, reject a destructive nature, or the violent communication of any new nature.

13. On account of the agreement and conformity of the similar effects which are wrought by heat and cold, reject motion of the body as a whole, whether expansive or contractive.

14. On account of heat being kindled by the attrition of bodies, reject a principial nature. By principial nature I mean that which exists in the nature of things positively, and not as the effect of any antecedent nature.

There are other natures beside these; for these tables are not perfect, but meant only for examples.

All and each of the above-mentioned natures do *not* belong to the form of heat. And from all of them man is freed in his operations of heat.

XIX

In the process of exclusion are laid the foundations of true induction, which however is not completed till it arrives at an affirmative. Nor is the exclusive part itself at all complete, nor indeed can it possibly be so at first. For exclusion is evidently the rejection of simple natures; and if we do not yet possess sound and true notions of simple natures, how can the process of exclusion be made accurate? Now some of the above-mentioned notions (as that of the nature of the elements, of the nature of heavenly bodies, of rarity) are vague and ill defined. I, therefore, well knowing and nowise forgetting how great a work I am about (viz., that of rendering the human understanding a match for things and nature), do not rest satisfied with the precepts I have laid down, but proceed further to devise and supply more powerful aids for the use of the understanding; which I shall now subjoin. And assuredly in the interpretation of nature the mind should by all means be so prepared and disposed that while it rests and finds footing in due stages and degrees of certainty, it may remember withal (especially at the beginning) that what it has before it depends in great measure upon what remains behind.

XX

And yet since truth will sooner come out from error than from confusion, I think it expedient that the understanding should have permission, after the three Tables of First Presentation (such as I have exhibited) have been made and weighed, to make an essay of the Interpretation of Nature in the affirmative way, on the strength both of the instances given in the tables, and of any others it may meet with elsewhere. Which kind of essay I call the *Indulgence of the Understanding,* or the *Commencement of Interpretation,* or the *First Vintage.*

First Vintage Concerning the Form of Heat

It is to be observed that the form of a thing is to be found (as plainly appears from what has been said) in each and all the instances in which the thing itself is to be found; otherwise it would not be the form. It follows therefore that there can be no contradictory instance. At the same time the form is found much more conspicuous and evident in some instances than in others, namely in those wherein the nature of the form is less restrained and obstructed and kept within bounds by other natures. Instances of this kind I call Shining or Striking Instances. Let us now therefore proceed to the first vintage concerning the form of heat.

From a survey of the instances, all and each, the nature of which heat is a particular case, appears to be motion. This is displayed most conspicuously in flame, which is always in motion, and in boiling or simmering liquids, which also are in perpetual motion. It is also shown in the excitement or increase of heat caused by motion, as in bellows and blasts; on which see Tab. 3. Inst. 29.; and again in other kinds of motion, on which see Tab. 3. Inst. 28. and 31. Again it is shown in the extinction of fire and heat by any strong compression, which checks and stops the motion; on which see Tab. 3. Inst. 30. and 32. It is shown also by this, that all bodies are destroyed, or at any rate notably altered, by all strong and vehement fire and heat; whence it is quite clear that heat causes a tumult and confusion and violent motion in the internal parts of a body, which perceptibly tend to its dissolution.

When I say of motion that it is as the genus of which heat is a species, I would be understood to mean not that heat generates motion or that motion generates heat (though both are true in certain cases), but that heat itself, its essence and quiddity, is motion and nothing else; limited however by the specific differences which I will presently subjoin, as soon as I have added a few cautions for the sake of avoiding ambiguity.

Sensible heat is a relative notion, and has relation to man, not to the universe, and is correctly defined as merely the effect of heat on the animal spirits. Moreover, in itself it is variable, since the same body, according as the senses are predisposed, induces a perception of cold as well as of heat. This is clear from Inst. 41. Tab. 3.

Nor again must the communication of heat, or its transitive nature, by means of which a body becomes hot when a hot body is applied to it, be confounded with the form of heat. For heat is one thing, heating another. Heat is produced by the motion of attrition without any preceding heat, an instance which excludes heating from the form of heat. And even when heat is produced by the approach of a hot body, this does not proceed from the form of heat, but depends entirely on a higher and more general nature, viz., on the nature of assimilation or self-multiplication, a subject which requires a separate inquiry.

Again, our notion of fire is popular, and of no use, being made up of the combination in any body of heat and brightness, as in common flame and bodies heated to redness.

Having thus removed all ambiguity, I come at length to the true specific differences which limit motion and constitute it the form of heat.

The first difference then is this. Heat is an expansive motion whereby a body strives to dilate and stretch itself to a larger sphere or dimension than it had previously occupied. This difference is most observable in flame, where the smoke or thick vapor manifestly dilates and expands itself into flame.

It is shown also in all boiling liquid which manifestly swells, rises, and bubbles, and carries on the process of self-expansion till it turns into a body far more extended and dilated than the liquid itself, namely, into vapor, smoke, or air.

It appears likewise in all wood and combustibles, from which there generally arises exudation and always evaporation.

It is shown also in the melting of metals which, being of the compactest texture, do not readily swell and dilate, but yet their spirit being dilated in itself, and thereupon conceiving an appetite for further dilation, forces and agitates the grosser parts into a liquid state. And if the heat be greatly increased it dissolves and turns much of their substance to a volatile state.

It is shown also in iron or stones which, though not melted or dissolved, are yet softened. This is the case also with sticks, which when slightly heated in hot ashes become flexible.

But this kind of motion is best seen in air, which continuously and manifestly dilates with a slight heat, as appears in Inst. 38. Tab. 3.

It is shown also in the opposite nature of cold. For cold contracts all bodies and makes them shrink, insomuch that in intense frosts nails fall out from walls, brazen vessels crack, and heated glass, on being suddenly placed in the cold, cracks and breaks. In like manner air is contracted by a slight chill, as in Inst. 38. Tab. 3. But on these points I shall speak more at length in the inquiry concerning Cold.

Nor is it surprising that heat and cold should exhibit many actions in common (for which see Inst. 32. Tab. 2.), when we find two of the following specific differences (of which I shall speak presently) suiting nature; though in this specific difference (of which I am now speaking) their actions are diametrically opposite. For heat gives an expansive and dilating, cold a contractive and condensing motion.

The second difference is a modification of the former, namely, that heat is a motion expansive or toward the circumference, but with this condition, that the body has at the same time a motion upward. For there is no doubt that there are many mixed motions. For instance, an arrow or dart turns as it goes forward, and goes forward as it turns. And in like manner the motion of heat is at once a motion of expansion and a motion upward. This difference is shown by putting a pair of tongs or a poker in the fire. If you put it in perpendicularly and hold it by the top, it soon burns your hand; if at the side or from below, not nearly so soon.

It is also observable in distillations *per descensorium,* which men use for delicate flowers that soon lose their scent. For human industry has discovered the plan of placing the fire not below but above, that it may burn the less. For not only flame tends upward, but also all heat.

But let trial be made of this in the opposite nature of cold, viz., whether cold does not contract a body downward as heat dilates a body upward. Take therefore two iron rods, or two glass tubes, exactly alike; warm them a little and place a sponge steeped in cold water or snow at the bottom of the one, and the same at the top of the other. For I think that the extremities of the rod which has the snow at the top will cool sooner than the extremities of the other which has the snow at the bottom; just as the opposite is the case with heat.

The third specific difference is this: that heat is a motion of expansion, not uniformly of the whole body together, but in the smaller parts of it; and at the same time checked, repelled, and beaten back, so that the body acquires a motion alternative, perpetually quivering, striving and struggling, and irritated by repercussion, whence springs the fury of fire and heat.

This specific difference is most displayed in flame and boiling liquids, which are perpetually quivering and swelling in small portions, and again subsiding.

It is also shown in those bodies which are so compact that when heated or ignited they do not swell or expand in bulk, as ignited iron, in which the heat is very sharp.

It is shown also in this, that a fire burns most briskly in the coldest weather.

Again, it is shown in this, that when the air is extended in a calendar glass without impediment or repulsion—that is to say, uniformly and equably—there is no perceptible heat. Also when wind escapes from confinement, although it burst forth with the greatest violence, there is no very great heat perceptible; because the motion is of the whole, without a motion alternating in the particles. And with a view to this, let trial be made whether flame does not burn more sharply toward the sides than in the middle of the flame.

It is also shown in this, that all burning acts on minute pores of the body burnt; so that burning undermines, penetrates, pricks, and stings the body like the points of an infinite number of needles. It is also an effect of this, that all strong waters (if suited to the body on which they are acting) act as fire does, in consequence of their corroding and pungent nature.

And this specific difference (of which I am now speaking)

is common also to the nature of cold. For in cold the contractive motion is checked by a resisting tendency to expand, just as in heat the expansive motion is checked by a resisting tendency to contract. Thus, whether the particles of a body work inward or outward, the mode of action is the same though the degree of strength be very different; because we have not here on the surface of the earth anything that is intensely cold. See Inst. 27. Tab. [1].

The fourth specific difference is a modification of the last: it is, that the preceding motion of stimulation or penetration must be somewhat rapid and not sluggish, and must proceed by particles, minute indeed, yet not the finest of all, but a degree larger.

This difference is shown by a comparison of the effects of fire with the effects of time or age. Age or time dries, consumes, undermines and reduces to ashes, no less than fire; indeed, with an action far more subtle; but because such motion is very sluggish, and acts on particles very small, the heat is not perceived.

It is also shown by comparing the dissolution of iron and gold. Gold is dissolved without any heat being excited, while the dissolution of iron is accompanied by a violent heat, though it takes place in about the same time. The reason is that in gold the separating acid enters gently and works with subtlety, and the parts of the gold yield easily; whereas in iron the entrance is rough and with conflict, and the parts of the iron have greater obstinacy.

It is shown also to some degree in some gangrenes and mortifications, which do not excite great heat or pain on account of the subtle nature of putrefaction.

Let this then be the First Vintage or Commencement of Interpretation concerning the form of heat, made by way of indulgence to the understanding.

Now from this our First Vintage it follows that the form or true definition of heat (heat, that is, in relation to the universe, not simply in relation to man) is, in few words, as follows: *Heat is a motion, expansive, restrained, and acting in its strife upon the smaller particles of bodies.* But the expansion is thus modified: *while it expands all ways, it has at the same time an inclination upward.* And the struggle in the particles is modified also; *it is not sluggish, but hurried and with violence.*

Viewed with reference to operation it is the same thing. For the direction is this: *If in any natural body you can excite a dilating or expanding motion, and can so repress this motion and turn it back upon itself that the dilation shall not proceed equably, but have its way in one part and be counteracted in another, you will undoubtedly generate heat;* without taking into account whether the body be elementary (as it is called) or subject to celestial influence; whether it be luminous or opaque; rare or dense; locally expanded or confined within the bounds of its first dimension; verging to dissolution or remaining in its original state; animal, vegetable, or mineral, water, oil or air, or any other substance whatever susceptible of the above-mentioned motion. Sensible heat is the same thing; only it must be considered with reference to the sense. Let us now proceed to further aids.

XXI

The Tables of First Presentation and the Rejection or process of Exclusion being completed, and also the First Vintage being made thereupon, we are to proceed to the other helps of the understanding in the Interpretation of Nature and true and perfect Induction. In propounding which, I mean, when Tables are necessary, to proceed upon the Instances of Heat and Cold; but when a smaller number of examples will suffice, I shall proceed at large; so that the inquiry may be kept clear, and yet more room be left for the exposition of the system.

I propose to treat, then, in the first place, of *Prerogative Instances;* secondly, of the *Supports of Induction;* thirdly, of the

Rectification of Induction; fourthly, of *Varying the Investigation according to the nature of the Subject;* fifthly, of *Prerogative Natures with respect to Investigation,* or of what should be inquired first and what last; sixthly, of the *Limits of Investigation,* or a synopsis of all natures in the universe; seventhly, of the *Application to Practice,* or of things in their relation to man; eighthly, of *Preparations for Investigation;* and lastly, of the *Ascending and Descending Scale of Axioms.*

XXII

Among Prerogative Instances I will place first *Solitary Instances.* Those are solitary instances which exhibit the nature under investigation in subjects which have nothing in common with other subjects except that nature; or, again, which do not exhibit the nature under investigation in subjects which resemble other subjects in every respect in not having that nature. For it is clear that such instances make the way short, and accelerate and strengthen the process of exclusion, so that a few of them are as good as many.

For instance, if we are inquiring into the nature of color, prisms, crystals, which show colors not only in themselves but externally on a wall, dews, etc., are solitary instances. For they have nothing in common with the colors fixed in flowers, colored stones, metals, woods, etc., except the color. From which we easily gather that color is nothing more than a modification of the image of light received upon the object, resulting in the former case from the different degrees of incidence, in the latter from the various textures and configurations of the body. These instances are solitary in respect to resemblance.

Again, in the same investigation, the distinct veins of white and black in marble, and the variegation of color in flowers of the same species, are solitary instances. For the black and white streaks in marble, or the spots of pink and white in a pink, agree in everything almost except the color. From which we easily gather that color has little to do with the intrinsic nature of a body, but simply depends on the coarser and as it were mechanical arrangement of the parts. These instances are solitary in respect to difference. Both kinds I call *solitary instances,* or *ferine,* to borrow a term from astronomers.

XXIII

Among Prerogative Instances I will next place *Migratory Instances.* They are those in which the nature in question is in the process of being produced when it did not previously exist, or on the other hand of disappearing when it existed before. And therefore, in either transition, such instances are always twofold, or rather it is one instance in motion or passage, continued till it reaches the opposite state. Such instances not only accelerate and strengthen the exclusive process, but also drive the affirmative or form itself into a narrow compass. For the form of a thing must necessarily be something which in the course of this migration is communicated, or on the other hand which in the course of this migration is removed and destroyed. And though every exclusion promotes the affirmative, yet this is done more decidedly when it occurs in the same than in different subjects. And the betrayal of the form in a single instance leads the way (as is evident from all that has been said) to the discovery of it in all. And the simpler the migration, the more must the instance be valued. Besides, migratory instances are of great use with a view to operation, because in exhibiting the form in connection with that which causes it to be or not to be, they supply a clear direction for practice in some cases; whence the passage is easy to the cases that lie next. There is, however, in these instances a danger which requires caution; viz., lest they lead us to connect the form too much with the efficient, and so possess the understanding, or at least touch it, with a false opinion concerning the form, drawn from a view of the efficient. But the efficient is always understood to be merely the vehicle that carries the form. This is a danger, however, easily remedied by the process of exclusion legitimately conducted.

I must now give an example of a migratory instance. Let the nature to be investigated be whiteness. An instance migrating to production or existence is glass whole and pounded. Again, simple water and water agitated into froth. For glass and water in their simple state are transparent, not white, whereas pounded glass and water in froth are white, not transparent. We must therefore inquire what has happened to the glass or water from this migration. For it is obvious that the form of whiteness is communicated and conveyed by that pounding of the glass and that agitation of the water. We find, however, that nothing has been added except the breaking up of the glass and water into small parts, and the introduction of air. But we have made no slight advance to the discovery of the form of whiteness when we know that two bodies, both transparent but in a greater or less degree (viz., air and water, or air and glass), do when mingled in small portions together exhibit whiteness, through the unequal refraction of the rays of light.

But an example must at the same time be given of the danger and caution to which I alluded. For at this point it might readily suggest itself to an understanding led astray by efficient causes of this kind, that air is always required for the form of whiteness, or that whiteness is generated by transparent bodies only—notions entirely false, and refuted by numerous exclusions. Whereas it will be found that (setting air and the like aside) bodies entirely even in the particles which affect vision are transparent, bodies simply uneven are white; bodies uneven and in a compound yet regular texture are all colors except black; while bodies uneven and in a compound, irregular, and confused texture are black. Here then I have given an example of an instance migrating to production or existence in the proposed nature of whiteness. An instance migrating to destruction in the same nature of whiteness is froth or snow in dissolution. For the water puts off whiteness and puts on transparency on returning to its integral state without air.

Nor must I by any means omit to mention that under migratory instances are to be included not only those which are passing toward production and destruction, but also those which are passing toward increase and decrease; since these also help to discover the form, as is clear from the above definition of form and the Table of Degrees. The paper, which is white when dry, but when wetted (that is, when air is excluded and water introduced) is less white and approaches nearer to the transparent, is analogous to the above given instances.

XXIV

Among Prerogative Instances I will put in the third place *Striking Instances,* of which I have made mention in the First Vintage Concerning Heat, and which I also call *Shining Instances,* or *Instances Freed and Predominant.* They are those which exhibit the nature in question naked and standing by itself, and also in its exaltation or highest degree of power; as being disenthralled and freed from all impediments, or at any rate by virtue of its strength dominant over, suppressing and coercing them. For since every body contains in itself many forms of natures united together in a concrete state, the result is that they severally crush, depress, break, and enthrall one another, and thus the individual forms are obscured. But certain subjects are found wherein the required nature appears more in its vigor than in others, either through the absence of impediments or the predominance of its own virtue. And instances of this kind strikingly display the form. At the same time in these instances also we must use caution, and check the hurry of the understanding. For whatever displays the form too conspicuously and seems to force it on the notice of the understanding should be held suspect, and recourse be had to a rigid and careful exclusion.

To take an example: let the nature inquired into be heat. A striking instance of the motion of expansion, which (as stated above) is the main element in the form of heat, is a calendar glass of air. For flame, though it manifestly exhibits expansion, still, as susceptible of momentary extinction, does not display the progress of expansion. Boiling water, too, on account of the easy transition of water to vapor or air, does not so well exhibit the expansion of water in its own body. Again, ignited iron and like bodies are so far from displaying the progress of expansion that in consequence of their spirit being crushed and broken by the coarse and compact particles which curb and subdue it, the expansion itself is not at all conspicuous to the senses. But a calendar glass strikingly displays expansion in air, at once conspicuous, progressive, permanent, and without transition.

To take another example: let the nature inquired into be weight. A striking instance of weight is quicksilver. For it far surpasses in weight all substances but gold, and gold itself is not much heavier. But quicksilver is a better instance for indicating the form of weight than gold, because gold is solid and consistent, characteristics which seem related to density; whereas quicksilver is liquid and teeming with spirit, and yet is heavier by many degrees than the diamond and other bodies that are esteemed the most solid. From which it is obvious that the form of heaviness or weight depends simply on quantity of matter and not on compactness of frame.

XXV

Among Prerogative Instances I will put in the fourth place *Clandestine Instances,* which I also call *Instances of the Twilight,* and which are pretty nearly the opposites of Striking Instances. For they exhibit the nature under investigation in its lowest degree of power, and as it were in its cradle and rudiments; striving indeed and making a sort of first attempt, but buried under and subdued by a contrary nature. Such instances, however, are of very great service for the discovery of forms; because as striking instances lead easily to specific differences, so are clandestine instances the best guides to *genera,* that is, to those common natures whereof the natures proposed are nothing more than particular cases.

For example, let the nature proposed be consistency, or the nature of that which determines its own figure, opposed to which is fluidity. Those are clandestine instances which exhibit some feeble and low degree of consistency in a fluid: as a bubble of water, which is a sort of consistent pellicle of determined figure, made of the body of the water. Of a similar kind are the droppings from a house, which if there be water to follow, lengthen themselves out into a very thin thread to preserve the continuity of the water; but if there be not water enough to follow, then they fall in round drops, which is the figure that best preserves the water from a solution of continuity. But at the very moment of time when the thread of water ceases and the descent in drops begins, the water itself recoils upward to avoid discontinuation. Again in metals, which in fusion are liquid but more tenacious, the molten drops often fly to the top and stick there. A somewhat similar instance is that of children's looking glasses, which little boys make on rushes with spittle, where also there is seen a consistent pellicle of water. This, however, is much better shown in that other childish sport when they take water, made a little more tenacious by soap, and blow it through a hollow reed, and so shape the water into a sort of castle of bubbles which by the interposition of the air become so consistent as to admit of being thrown some distance without discontinuation. But best of all is it seen in frost and snow, which assume such a consistency that they can be almost cut with a knife, although they are formed out of air and water, both fluids. All which facts not obscurely intimate that consistent and fluid are only vulgar notions, and relative to the sense; and that in fact there is inherent in all bodies a disposition to shun and escape discontinuation; but that it is faint and feeble in homogeneous bodies (as fluids), more lively and strong in bodies compounded of heterogeneous matter; the reason being that the approach of heterogeneous matter binds bodies together, while the insinuation of homogeneous matter dissolves and relaxes them.

To take another instance, let the proposed nature be the attraction or coming together of bodies. In the investigation of its form the most remarkable striking instance is the magnet. But there is a contrary nature to the attractive; namely, the nonattractive, which exists in a similar substance. Thus there is iron which does not attract iron, just as lead does not attract lead, nor wood wood, nor water water. Now a clandestine instance is a magnet armed with iron, or rather the iron in an armed magnet. For it is a fact in nature that an armed magnet at some distance off does not attract iron more powerfully than an unarmed magnet. But if the iron be brought so near as to touch the iron in the armed magnet, then the armed magnet supports a far greater weight of iron than a simple and unarmed magnet, on account of the similarity of substance between the pieces of iron; an operation altogether clandestine and latent in the iron before the magnet was applied. Hence it is manifest that the form of coition is something which is lively and strong in the magnet, feeble and latent in iron. Again, it has been observed that small wooden arrows without an iron point, discharged from large engines, pierce deeper into wooden material (say the sides of ships, or the like) than the same arrows tipped with iron, on account of the similarity of substance between the two pieces of wood; although this property had previously been latent in the wood. In like manner, although air does not manifestly attract air or water water in entire bodies, yet a bubble is more easily dissolved on the approach of another bubble than if that other bubble were away, by reason of the appetite of coition between water and water, and between air and air. Such clandestine instances (which, as I have said, are of the most signal use) exhibit themselves most conspicuously in small and subtle portions of bodies; the reason being that larger masses follow more general forms, as shall be shown in the proper place.

XXVI

Among Prerogative Instances I will put in the fifth place *Constitutive Instances,* which I also call *Manipular.* They are those which constitute a single species of the proposed nature, a sort of Lesser Form. For since the genuine forms (which are always convertible with the proposed natures) lie deep and are hard to find, it is required by the circumstances of the case and the infirmity of the human understanding that particular forms, which collect together certain groups of instances (though not all) into some common notion, be not neglected, but rather be diligently observed. For whatever unites nature, though imperfectly, paves the way to the discovery of forms. Instances, therefore, which are useful in this regard are of no despicable power, but have a certain prerogative.

But great caution must here be employed lest the human understanding, after having discovered many of those particular forms and thereupon established partitions or divisions of the nature in question, be content to rest therein, and instead of proceeding to the legitimate discovery of the great form, take it for granted that the nature from its very roots is manifold and divided, and so reject and put aside any further union of the nature, as a thing of superfluous subtlety and verging on mere abstraction.

For example, let the proposed nature be memory, or that which excites and aids the memory. Constitutive instances are: order or distribution, which clearly aids the memory; also topics or "places" in artificial memory; which may either be *places* in the proper sense of the word, as a door, angle, window, and the like; or familiar and known persons; or any other things at pleasure (provided they be placed in a certain order), as animals, vegetables; words, too, letters, characters, historical persons, and the like; although some of these are more suitable and convenient than others. Such artificial places help the memory wonderfully, and exalt it far above its natural powers. Again, verse is learned and remembered more easily than prose. From this group of three instances, viz., order, artificial places, and verse, one species of aid to the memory is constituted. And this species may with propriety be called the cutting off of infinity. For when we try to recollect or call a thing to mind, if we have no prenotion or perception of what we are seeking, we seek and toil and wander here and there, as if in infinite space. Whereas, if we have any sure pre-notion, infinity is at once cut off, and the memory has not so far to range. Now in the three foregoing instances the preno-tion is clear and certain. In the first it must be something which suits the order; in the second it must be an image which bears some relation or conformity to the places fixed; in the third, it must be words that fall into the verse; and thus infinity is cut off. Other instances, again, will give us this second species: that whatever brings the intellectual con-ception into contact with the sense (which is indeed the method most used in mnemonics) assists the memory. Other instances will give us this third species: that things which make their impression by way of a strong affection, as by in-spiring fear, admiration, shame, delight, assist the memory. Other instances will give us this fourth species: that things which are chiefly imprinted when the mind is clear and not occupied with anything else either before or after, as what is learned in childhood, or what we think of before going to sleep, also things that happen for the first time, dwell longest in the memory. Other instances will give us this fifth species: that a multitude of circumstances or points to take hold of aids the memory; as writing with breaks and divisions, read-ing or reciting aloud. Lastly, other instances will give us this sixth species: that things which are waited for and raise the attention dwell longer in the memory than what flies quickly by. Thus, if you read anything over twenty times, you will not learn it by heart so easily as if you were to read it only ten, trying to repeat it between whiles, and when memory failed, looking at the book. It appears, then, that there are six lesser forms of aids to the memory; viz.: the cutting off of infinity; the reduction of the intellectual to the sensible; impression made on the mind in a state of strong emotion; impression made on the mind disengaged; multitude of points to take hold of; expectation beforehand.

To take another example, let the proposed nature be taste or tasting. The following instances are Constitutive. Persons who are by nature without the sense of smell cannot perceive or distinguish by taste food that is rancid or putrid, nor food that is seasoned with garlic, or with roses, or the like. Again, persons whose nostrils are accidentally obstructed by a catarrh cannot distinguish or perceive anything putrid or rancid or sprinkled with rosewater. Again, persons thus affected with catarrh, if while they have something fetid or perfumed in their mouth or palate they blow their nose violently, immediately perceive the rancidity or the perfume. These instances, then, will give and constitute this species, or rather division, of taste: that the sense of taste is in part nothing else than an internal smell, passing and descending from the upper passages of the nose to the mouth and palate. On the other hand the tastes of salt, sweet, sour, acid, rough, bitter, and the like, are as perceptible to those in whom the sense of smell is wanting or stopped as to anyone else; so that it is clear that the sense of taste is a sort of compound of an internal smell and a delicate power of touch—of which this is not the place to speak.

To take another example, let the proposed nature be the communication of quality without admixture of substance. The instance of light will give or constitute one species of communication; heat and the magnet another. For the communication of light is momentaneous, and ceases at once on the removal of the original light. But heat and the virtue of the magnet, after they have been transmitted to or rather excited in a body, lodge and remain there for a considerable time after the removal of the source of motion.

Very great, in short, is the prerogative of constitutive instances; for they are of much use in the forming of definitions (especially particular definitions) and in the division and partition of natures; with regard to which it was not ill said by Plato, "That he is to be held as a god who knows well how to define and to divide."

XXVII

Among Prerogative Instances I will put in the sixth place *Instances Conformable,* or *of Analogy;* which I also call *Parallels,* or *Physical Resemblances.* They are those which represent the resemblances and conjugations of things, not in lesser forms (as constitutive instances do) but merely in the concrete. Hence they may be called the first and lowest steps toward the union of nature. Nor do they constitute any axiom immediately from the beginning, but simply point out and mark a certain agreement in bodies. But although they are of little use for the discovery of forms, they nevertheless are very serviceable in revealing the fabric of the parts of the universe, and anatomizing its members; from which they often lead us along to sublime and noble axioms, especially those which relate to the configuration of the world rather than to simple forms and natures.

For example, these following are instances of conformity: a looking glass and the eye; and again, the construction of the ear and places returning an echo. From which conformity, to say nothing of the mere observation of the resemblance which is in many respects useful, it is easy to gather and form this axiom—that the organs of the senses, and bodies which produce reflections to the senses, are of a like nature. Again, upon this hint the understanding easily rises to a higher and nobler axiom, which is this: that there is no difference between the consents or sympathies of bodies endowed with sensation and those of inanimate bodies without sensation, except that in the former an animal spirit is added to the body so disposed, but is wanting in the latter. Whence it follows that there might be as many senses in animals as there are sympathies between inanimate bodies, if there were perforations in the animate body allowing the animal spirit to pass freely into a member rightly disposed, as into a fit organ. Again, as many as are the senses in animals, so many without doubt are the motions in an inanimate body where animal spirit is wanting; though necessarily there are many more motions in inanimate bodies than there are senses in animate, on account of the paucity of organs of sense. And of this a manifest example is exhibited in pain. For though there are many kinds and varieties of pain in animals (as the pain of burning, for one, of intense cold for another; again, of pricking, squeezing, stretching, and the like), it is yet most certain that all of them, as far as the motion is concerned, exist in inanimate substances; for example, in wood or stone, when it is burned or frozen or pricked or cut or bent or stretched, and so on, though they do not enter the senses for want of the animal spirit.

Again, the roots and branches of plants (which may seem strange) are conformable instances. For all vegetable matter swells and pushes out its parts to the surface, as well upward as downward. Nor is there any other difference between roots and branches than that the root is buried in the ground, while the branches are exposed to the air and sun. For if you take a tender and flourishing branch of a tree, and bend it down into a clod of earth, although it does not cohere with the ground itself, it presently produces not a branch but a root. And vice versa, if earth be placed at the top, and so kept down with a stone or any hard substance as to check the plant and prevent it from shooting upward, it will put forth branches into the air downward.

Again, the gums of trees, and most rock gems, are conformable instances. For both of these are nothing else than exudations and filterings of juices, the former from trees, the latter from rocks; whence is produced the splendor and clearness in each, that is, by the fine and delicate filtering. Hence, too, it is that the hairs of animals are not generally so beautiful and of so vivid a color as the feathers of birds, viz., because the juices do not filter so finely through skin as through quills.

Again, the scrotum in males and the matrix in females are conformable instances. So that the great organic difference between the sexes (in land animals at least) appears to be nothing more than that the one organization is external and the other internal. That is to say, the greater force of heat in the male thrusts the genitals outward; whereas in the female the heat is too feeble to effect this, and thus they are contained within.

The fins of fish, again, and the feet of quadrupeds, or the feet and wings of birds, are conformable instances; to which Aristotle has added the four folds in the motions of serpents. Whence it appears that in the structure of the universe the motions of living creatures are generally effected by a quaternion of limbs or of bendings.

Again, the teeth of land animals and the beaks of birds are conformable instances; from which it is manifest that in all perfect animals there is a determination of some hard substance to the mouth.

Nor is that an absurd similitude of conformity which has been remarked between man and a plant inverted. For the root of the nerves and faculties in animals is the head, while the seminal parts are the lowest—the extremities of the legs and arms not reckoned. In a plant, on the other hand, the root (which answers to the head) is regularly placed in the lowest part, and the seeds in the highest.

To conclude, it cannot too often be recommended and enjoined that men's diligence in investigating and amassing natural history be henceforward entirely changed and turned into the direction opposite to that now in use. For hitherto men have used great and indeed overcurious diligence in observing the variety of things, and explaining the exact specific differences of animals, herbs, and fossils; most of which are rather sports of nature than of any serious use toward science. Such things indeed serve to delight, and sometimes even give help in practice; but for getting insight into nature they are of little service or none. Men's labor therefore should be turned to the investigation and observation of the resemblances and analogies of things, as well in wholes as in parts. For these it is that detect the unity of nature, and lay a foundation for the constitution of sciences.

But here must be added a strict and earnest caution, that those only are to be taken for conformable and analogous instances which indicate (as I said at the beginning) physical resemblances, that is, real and substantial resemblances; resemblances grounded in nature, not accidental or merely apparent; much less superstitious or curious resemblances, such as the writers on natural magic (very frivolous persons, hardly to be named in connection with such serious matters as we are now about) are everywhere parading—similitudes and sympathies of things that have no reality, which they describe and sometimes invent with great vanity and folly.

But to leave these. The very configuration of the world itself in its greater parts presents conformable instances which are not to be neglected. Take, for example, Africa and the region of Peru with the continent stretching to the Straits of Magellan, in each of which tracts there are similar isthmuses and similar promontories, which can hardly be by accident.

Again, there is the Old and New World, both of which are broad and extended towards the north, narrow and pointed towards the south.

We have also most remarkable instances of conformity in the intense cold existing in what is called the middle region of the air and the violent fires which are often found bursting forth from beneath the ground, which two things are *ultimities* and extremes; that is to say, the extreme of the nature of cold toward the circumference of the sky, of heat toward the bowels of the earth, by *antiperistasis* or the rejection of the contrary nature.

Lastly, the conformity of instances in the axioms of science is deserving of notice. Thus the rhetorical trope of deceiving expectation is conformable with the musical trope of avoiding or sliding from the close or cadence; the mathematical postulate that if two things are equal to the same thing they are equal to one another is conformable with the rule of the syllogism in logic which unites propositions agreeing in a middle term. In fine, a certain sagacity in investigating and hunting out physical conformities and similitudes is of very great use in very many cases.

XXVIII

Among Prerogative Instances I will put in the seventh place *Singular Instances,* which I also call *Irregular* or *Heteroclite,* to borrow a term from grammarians. They are such as exhibit bodies in the concrete, which seem to be out of the course and broken off from the order of nature, and not agreeing with other bodies of the same kind. For conformable instances are like each other; singular instances are like themselves alone. The use of singular instances is the same as that of clandestine, namely, to raise and unite nature for the purpose of discovering kinds of common natures, to be afterward limited by true specific differences. For we are not to give up the investigation until the properties and qualities found in such things as may be taken for miracles of nature be reduced and comprehended under some form or fixed law, so that all the irregularity or singularity shall be found to depend on some common form, and the miracle shall turn out to be only in the exact specific differences, and the degree, and the rare concurrence, not in the species itself. Whereas now the thoughts of men go no further than to pronounce such things the secrets and mighty works of nature, things as it were causeless, and exceptions to general rules.

Examples of singular instances are the sun and moon among stars; the magnet among stones; quicksilver among metals; the elephant among quadrupeds; the venereal sense among kinds of touch; the scent of hounds among kinds of smell. So among grammarians the letter S is held singular, on account of its easy combination with consonants, sometimes with two, sometimes even with three, which property no other letter has. Such instances must be regarded as most valuable, because they sharpen and quicken investigation and help to cure the understanding depraved by custom and the common course of things.

XXIX

Among Prerogative Instances I will put in the eighth place *Deviating Instances,* that is, errors, vagaries, and prodigies of nature, wherein nature deviates and turns aside from her ordinary course. Errors of nature differ from singular instances in this, that the latter are prodigies of species, the former of individuals. Their use is pretty nearly the same, for they correct the erroneous impressions suggested to the understanding by ordinary phenomena, and reveal common forms. For in these also we are not to desist from inquiry until the cause of the deviation is discovered. This cause, however, does not rise properly to any form, but simply to the latent process that leads to the form. For he that knows the ways of nature will more easily observe her deviations; and on the other hand he that knows her deviations will more accurately describe her ways.

They differ in this also from singular instances, that they give much more help to practice and the operative part. For to produce new species would be very difficult, but to vary known species and thereby produce many rare and unusual results is less difficult. Now it is an easy passage from miracles of nature to miracles of art. For if nature be once detected in her deviation, and the reason thereof made evident, there will be little difficulty in leading her back by art to the point whither she strayed by accident; and that not only in one case, but also in others. For errors on one side point out and open the way to errors and deflections on all sides. Under this head there is no need of examples, they are so plentiful. For we have to make a collection or particular natural history of all prodigies and monstrous births of nature; of everything in short that is in nature new, rare, and unusual. This must be done, however, with the strictest scrutiny, that fidelity may be ensured. Now those things are to be chiefly suspected which depend in any way on religion, as the prodigies of Livy, and those not less which are found in writers on natural magic or alchemy, and men of that sort, who are a kind of suitors and lovers of fables. But whatever is admitted must be drawn from grave and credible history and trustworthy reports.

XXX

Among Prerogative Instances I will put in the ninth place *Bordering Instances*, which I also call *Participles*. They are those which exhibit species of bodies that seem to be composed of two species, or to be rudiments between one species and another. These instances might with propriety be reckoned among singular or heteroclite instances, for in the whole extent of nature they are of rare and extraordinary occurrence. But nevertheless for their worth's sake they should be ranked and treated separately, for they are of excellent use in indicating the composition and structure of things, and suggesting the causes of the number and quality of the ordinary species in the universe, and carrying on the understanding from that which is to that which may be.

Examples of these are: moss, which holds a place between putrescence and a plant; some comets, between stars and fiery meteors; flying fish, between birds and fish; bats, between birds and quadrupeds; also the ape, between man and beast—

Simia quam similis turpissima bestia nobis;

likewise the biformed births of animals, mixed of different species, and the like.

XXXI

Among Prerogative Instances I will put in the tenth place *Instances of Power,* or of the *Fasces* (to borrow a term from the badges of empire); which I also call *Instances of the Wit,* or *Hands of Man*. These are the noblest and most consummate works in each art, exhibiting the ultimate perfection of it. For since our main object is to make nature serve the business and conveniences of man, it is altogether agreeable to that object that the works which are already in man's power should (like so many provinces formerly occupied and subdued) be noted and enumerated, especially such as are the most complete and perfect; because starting from them we shall find an easier and nearer passage to new works hitherto unattempted. For if from an attentive contemplation of these a man pushes on his work with zeal and activity, he will infallibly either advance them a little further, or turn them aside to something in their neighborhood, or even apply and transfer them to some more noble use.

Nor is this all. But as by rare and extraordinary works of nature the understanding is excited and raised to the investigation and discovery of forms capable of including them, so also is this done by excellent and wonderful works of art, and that in a much greater degree, because the method of creating and constructing such miracles of art is in most cases plain, whereas in the miracles of nature it is generally obscure. But with these also we must use the utmost caution lest they depress the understanding and fasten it as it were to the ground.

For there is danger lest the contemplation of such works of art, which appear to be the very summits and crowning points of human industry, may so astonish and bind and bewitch the understanding with regard to them, that it shall be incapable of dealing with any other, but shall think that nothing can be done in that kind except by the same way in which these were done—only with the use of greater diligence and more accurate preparation.

Whereas on the contrary this is certain: that the ways and means of achieving the effects and works hitherto discovered and observed are for the most part very poor things, and that all power of a high order depends on forms and is derived in order from the sources thereof; not one of which has yet been discovered.

And therefore (as I have said elsewhere) if a man had been thinking of the war engines and battering-rams of the ancients, though he had done it with all his might and spent his whole life in it, yet he would never have lighted on the discovery of cannon acting by means of gunpowder. Nor again, if he had fixed his observation and thought on the manufacture of wool and cotton, would he ever by such means have discovered the nature of the silkworm or of silk.

Hence it is that all the discoveries which can take rank among the nobler of their kind have (if you observe) been brought to light, not by small elaborations and extensions of arts, but entirely by accident. Now there is nothing which can forestall or anticipate accident (which commonly acts only at long intervals) except the discovery of forms.

Particular examples of such instances it is unnecessary to adduce, for there is such an abundance of them. For what we have to do is simply this: to seek out and thoroughly inspect all mechanical arts, and all liberal too (as far as they deal with works), and make therefrom a collection or particular history of the great and masterly and most perfect works in every one of them, together with the mode of their production or operation.

And yet I do not tie down the diligence that should be used in such a collection to those works only which are esteemed the masterpieces and mysteries of any art, and which excite wonder. For wonder is the child of rarity; and if a thing be rare, though in kind it be no way extraordinary, yet it is wondered at. While on the other hand things which really call for wonder on account of the difference in species which they exhibit as compared with other species, yet if we have them by us in common use, are but slightly noticed.

Now the singularities of art deserve to be noticed no less than those of nature, of which I have already spoken. And as among the singularities of nature I placed the sun, the moon, the magnet, and the like—things in fact most familiar, but in nature almost unique—so also must we do with the singularities of art.

For example, a singular instance of art is paper, a thing exceedingly common. Now if you observe them with attention, you will find that artificial materials are either woven in upright and transverse threads, as silk, woolen or linen cloth, and the like; or cemented of concreted juices, as brick, earthenware, glass, enamel, porcelain, etc., which are bright if well united, but if not, are hard indeed but not bright. But all things that are made of concrete juices are brittle, and no way cohesive or tenacious. On the contrary, paper is a tenacious substance that may be cut or torn; so that it imitates and almost rivals the skin or membrane of an animal, the leaf of a vegetable, and the like pieces of nature's workmanship. For it is neither brittle like glass, nor woven as cloth; but is in fibers, not distinct threads, just like natural materials; so that among artificial materials you will hardly find anything similar; but it is altogether singular. And certainly among things artificial those are to be preferred which either come nearest to an imitation of nature, or on the contrary overrule and turn her back.

Again, as instances of the wit and hand of man, we must not altogether contemn juggling and conjuring tricks. For some of them, though in use trivial and ludicrous, yet in regard to the information they give may be of much value.

Lastly, matters of superstition and magic (in the common acceptation of the word) must not be entirely omitted. For although such things lie buried deep beneath a mass of falsehood and fable, yet they should be looked into a little. For it may be that in some of them some natural operation lies at the bottom, as in fascination, strengthening of the imagination, sympathy of things at a distance, transmission of impressions from spirit to spirit no less than from body to body, and the like.

XXXII

From what has been said it is clear that the five classes of instances last mentioned (namely, Instances Conformable, Singular, Deviating, Bordering, and of Power) ought not to be reserved until some certain nature be in question (as the other instances which I have placed first, and most of those that are to follow should), but a collection of them must be begun at once, as a sort of particular history; because they serve to digest the matters that enter the understanding, and to correct the ill complexion of the understanding itself, which cannot but be tinged and infected, and at length perverted and distorted, by daily and habitual impression.

These instances therefore should be employed as a sort of preparative for setting right and purging the understanding. For whatever withdraws the understanding from the things to which it is accustomed, smooths and levels its surface for the reception of the dry and pure light of true ideas.

Moreover such instances pave and prepare the way for the operative part, as will be shown in the proper place, when I come to speak of deductions leading to Practice.

XXXIII

Among Prerogative Instances I will put in the eleventh place *Instances of Companionship and of Enmity*, which I also call Instances of *Fixed Propositions*. They are those which exhibit a body or concrete substance in which the nature inquired into constantly attends, as an inseparable companion; or in which on the contrary it constantly retreats, and is excluded from companionship as an enemy and foe. For from such instances are formed certain and universal propositions, either affirmative or negative, in which the subject will be a body in concrete, and the predicate the nature itself that is in question. For particular propositions are in no case fixed. I mean propositions in which the nature in question is found in any concrete body to be fleeting and movable, that is to say accruing or acquired, or on the other hand departing or put away. Wherefore particular propositions have no prerogative above others, save only in the case of migration, of which I have already spoken. Nevertheless even these particular propositions being prepared and collated with universal propositions are of great use, as shall be shown in the proper place. Nor even in the universal propositions do we require exact or absolute affirmation or negation. For it is sufficient for the purpose in hand even if they admit of some rare and singular exception.

The use of instances of companionship is to bring the affirmative of the form within narrow limits. For if by migratory instances the affirmative of the form is narrowed to this, that the form of the thing must needs be something which by the act of migration is communicated or destroyed; so in instances of companionship, the affirmative of the form is narrowed to this, that the form of the thing must needs be something which enters as an element into such a concretion of body, or contrariwise which refuses to enter; so that he who well knows the constitution or configuration of such a body will not be far from bringing to light the form of the nature under inquiry.

For example, let the nature in question be heat. An instance of companionship is flame. For in water, air, stone, metal, and most other substances, heat is variable, and may come and go, but all flame is hot, so that heat is always in attendance on the concretion of flame. But no hostile instance of heat is to be found here. For the senses know nothing of the bowels of the earth, and of all the bodies which we do know there is not a single concretion that is not susceptible to heat.

But to take another instance: let the nature in question be consistency. A hostile instance is air. For metal can be fluid and can also be consistent; and so can glass; water also can be consistent, when it is frozen; but it is impossible that air should ever be consistent, or put off its fluidity.

But with regard to such instances of fixed propositions I

have two admonitions to give, which may help the business in hand. The first is that, if a universal affirmative or negative be wanting, that very thing be carefully noted as a thing that is not; as we have done in the case of heat, where a universal negative (as far as the essences that have come under our knowledge are concerned) is not to be found in the nature of things. In like manner, if the nature in question be eternity or incorruptibility, no universal affirmative is to be found here. For eternity or incorruptibility cannot be predicated of any of the bodies lying below the heavens and above the bowels of the earth. The other admonition is that to universal propositions, affirmative or negative, concerning any concrete body, there be subjoined those concretes which seem to approach most nearly to that which is not; as in heat, the gentlest and least burning flames; in incorruptibility, gold which comes nearest to it. For all such indicate the limits of nature between that which is and that which is not, and help to circumscribe forms and prevent them from escaping and straying beyond the conditions of matter.

XXXIV

Among Prerogative Instances I will put in the twelfth place those *Subjunctive Instances* mentioned in the last aphorism, which I otherwise call *Instances of Ultimity or Limit*. For such instances are not only useful when subjoined to fixed propositions, but also by themselves and in their own properties. For they point out not obscurely the real divisions of nature and measures of things, and how far in any case nature may act or be acted upon, and then the passages of nature into something else. Of this kind are gold in weight; iron in hardness; the whale in animal bulk; the dog in scent; the combustion of gunpowder in rapid expansion; and the like. Nor should extremes in the lowest degree be less noticed than extremes in the highest; such as spirit of wine in weight; silk in softness; the worms of the skin in animal bulk; and the like.

XXXV

Among Prerogative Instances I will put in the thirteenth place *Instances of Alliance or Union*. They are those which mingle and unite natures supposed to be heterogeneous, and marked and set down as such in the received divisions.

Instances of alliance show that operations and effects attributed to some one heterogeneous nature as peculiar to it may belong also to other heterogeneous natures; that this supposed heterogeneity is proved to be not real or essential, but only a modification of a common nature. They are therefore of most excellent use in raising and elevating the understanding from specific differences to *genera,* and in dispelling phantoms and false images of things, which in concrete substances come before us in disguise. For example, let the nature in question be heat. We are told (and it seems to be a division quite received and authorized) that there are three kinds of heat: the heat of heavenly bodies, the heat of animals, and the heat of fire; and that these heats (especially one of them as compared with the other two) are in their very essence and species—that is to say, in their specific nature—distinct and heterogeneous, since the heat of heavenly bodies and of animals generates and cherishes, while the heat of fire wastes and destroys. We have, therefore, an instance of alliance in that common case, when the branch of a vine is brought within a house where a fire is constantly kept up, and the grapes ripen on it a whole month sooner than they do out of doors; so that the ripening of fruit, even while it hangs on the tree, may be brought about by fire, though such ripening would seem to be the proper work of the sun. From this beginning, therefore, the understanding, rejecting the notion of essential heterogeneity, easily rises to inquire what are in reality those points of difference between the heat of the sun and of fire which cause their operations to be so dissimilar, however they may themselves partake of a common nature.

These differences will be found to be four. The first is

that the heat of the sun compared with the heat of fire is far milder and softer in degree; the second is that in quality (at least as it reaches us through the air) it is far moister; the third (and this is the main point) is that it is exceedingly unequal, now approaching and increased, now receding and diminished; which thing chiefly contributes to the generation of bodies. For Aristotle was right in asserting that the principal cause of the generations and corruptions which are going on here on the surface of the earth is the oblique course of the sun through the zodiac; whence the heat of the sun, partly by the alternation of day and night, partly by the succession of summer and winter, becomes strangely unequal. And yet this great man must go on at once to corrupt and deprave what he has rightly discovered. For laying down the law to nature (as his way is), he very dictatorially assigns as the cause of generation the approach of the sun, and as the cause of corruption his retreat; whereas both together (the approach of the sun and his retreat), not respectively, but as it were indifferently, afford a cause both for generation and production; since inequality of heat ministers to generation and corruption, equality to conservation only. There is also a fourth specific difference between the heat of the sun and of fire, and one of very great moment; viz., that the sun operates by gentle action through long spaces of time, whereas the operations of fire, urged on by the impatience of man, are made to finish their work in shorter periods. But if anyone were to set to work diligently to temper the heat of fire and reduce it to a milder and more moderate degree, as is easily done in many ways, and were then to sprinkle and intermix a little moisture; and if above all he were to imitate the heat of the sun in its inequality; and lastly if he could submit to a slow procedure, not indeed corresponding to the operations of the sun, but yet slower than men generally adopt in working with fire; he would speedily get rid of the notion of different kinds of heat, and would attempt to imitate, if not equal or in some cases even surpass the works of the sun by the heat of fire. We have a similar instance of alliance in the revival of butterflies stupefied and half dead with cold, by slightly warming them at a fire. So that you may easily see that fire is no more without the power of giving life to animals than of ripening vegetables. Thus also Fracastorius' celebrated invention of the heated pan with which doctors cover the heads of apoplectic patients who are given over, manifestly expands the animal spirits, compressed and all but extinguished by the humors and obstructions of the brain, and exciting them to motion, just as fire acts on air or water, by consequence quickens and gives them life. Eggs also are sometimes hatched by the heat of fire, which thus exactly imitates animal heat. And there are many instances of the same kind, so that no one can doubt that the heat of fire may in many subjects be modified so as to resemble the heat of heavenly bodies and of animals.

Again, let the natures in question be motion and rest. It appears to be a received division and drawn from the depths of philosophy, that natural bodies either move in circle, or move straight forward, or remain at rest. For there is either motion without limit, or rest at a limit, or progress toward a limit. Now, that perpetual motion of rotation seems to be proper to the heavenly bodies, station or rest seems to belong to the globe of the earth, while other bodies (which they call heavy or light, being indeed placed out of the region to which they naturally belong) are carried toward the masses or congregations of their likes; light bodies upward toward the circumference of the heaven, heavy bodies downward towards the earth. And this is pretty talk.

But we have an instance of alliance in one of the lower comets, which though far below the heaven, nevertheless revolve. And Aristotle's fiction of a comet being tied to or following some particular star has long been exploded, not only because the reason for it is not probable, but because we have manifest experience of the discursive and irregular motion of comets through various parts of the sky.

Again, another instance of alliance on this subject is the motion of air, which within the tropics, where the circles of rotation are larger, seems itself also to revolve from east to west.

Again, another instance would be the ebb and flow of the sea, if it be found that the waters themselves are carried in a motion of rotation (however slow and evanescent) from east to west, though subject to the condition of being driven back twice in the day. For if things be so, it is manifest that that motion of rotation is not limited to heavenly bodies, but is shared also by air and water.

Even that property of light substances, viz., that they tend upward, is somewhat at fault. And on this point a bubble of water may be taken as an instance of alliance. For if there be air under the water it rapidly ascends to the surface by that motion of percussion (as Democritus calls it) by which the descending water strikes and raises the air upward; not by any effort or struggle of the air itself. And when it is come to the surface of the water, then the air is stopped from further ascent by a slight resistance it meets with in the water, which does not immediately allow itself to be separated; so that the desire of air to ascend must be very slight.

Again, let the nature in question be weight. It is quite a received division that dense and solid bodies move toward the center of the earth, rare and light toward the circumference of the heaven, as to their proper places. Now as for this notion of places, though such things prevail in the schools, it is very silly and childish to suppose that place has any power. Therefore philosophers do but trifle when they say that if the earth were bored through, heavy bodies would stop on reaching the center. Certainly it would be a wonderful and efficacious sort of nothing, or mathematical point, which could act on bodies, or for which bodies could have desire, for bodies are not acted on except by bodies. But this desire of ascending and descending depends either on the configuration of the body moved or on its sympathy or consent with some other body. Now if there be found any body which, being dense and solid, does not move to the earth, there is an end of this division. But if

Gilbert's opinion be received, that the earth's magnetic power of attracting heavy bodies does not extend beyond the orb of its virtue (which acts always to a certain distance and no more), and if this opinion be verified by a single instance, in that we shall have got at last an instance of alliance on the subject of weight. But at present there does not occur any instance on this subject certain and manifest. What seems to come nearest to one is that of the waterspouts, often seen in the voyage over the Atlantic Ocean toward either of the Indies. For so great is the quantity and mass of water suddenly discharged by these waterspouts that they seem to have been collections of water made before, and to have remained hanging in these places, and afterward to have been rather thrown down by some violent cause, than to have fallen by the natural motion of gravity. So that it may be conjectured that a dense and compact mass, at a great distance from the earth, would hang like the earth itself and not fall unless thrust down. But on this point I affirm nothing certain. Meanwhile in this and many other cases it will easily be seen how poor we are in natural history, when in place of certain instances I am sometimes compelled to adduce as examples bare suppositions.

Again, let the nature in question be discourse of reason. The distinction between human reason and the sagacity of brutes appears to be a perfectly correct one. Yet there are certain instances of actions performed by animals, by which it seems that brutes too have some power of syllogizing; as in the old story of the crow which, in a time of great drought being half dead with thirst, saw some water in the hollow trunk of a tree, and finding it too narrow to get in, proceeded to drop in a number of pebbles till the water rose high enough for it to drink; and this afterward passed into a proverb.

Again, let the nature in question be visibility. It appears to be a very correct and safe division which regards light as primarily visible, and affording the power of seeing; while color is secondarily visible, and cannot be seen without light, so that it appears to be nothing more than an image or modification of light. And yet there appear to be instances of alliance on either side, namely, snow in great quantities, and the flame of sulphur; in one of which there appears to be a color primarily giving light, in the other a light verging on color.

XXXVI

Among Prerogative Instances I will put in the fourteenth place *Instances of the Fingerpost,* borrowing the term from the fingerposts which are set up where roads part, to indicate the several directions. These I also call *Decisive and Judicial,* and in some cases, *Oracular* and *Commanding Instances.* I explain them thus. When in the investigation of any nature the understanding is so balanced as to be uncertain to which of two or more natures the cause of the nature in question should be assigned on account of the frequent and ordinary concurrence of many natures, instances of the fingerpost show the union of one of the natures with the nature in question to be sure and indissoluble, of the other to be varied and separable; and thus the question is decided, and the former nature is admitted as the cause, while the latter is dismissed and rejected. Such instances afford very great light and are of high authority, the course of interpretation sometimes ending in them and being completed. Sometimes these instances of the fingerpost meet us accidentally among those already noticed, but for the most part they are new, and are expressly and designedly sought for and applied, and discovered only by earnest and active diligence.

For example, let the nature in question be the ebb and flow of the sea; each of which is repeated twice a day, and takes six hours each time, subject to some slight difference which coincides with the motion of the moon. The following will be a case of the parting of the roads.

This motion must necessarily be caused either by the advance and retreat of the waters, as water shaken in a basin leaves one side when it washes the other; or else by a lifting up of the waters from the bottom and falling again, as water in boiling rises and falls. The question is to which of these two causes the ebb and flow should be assigned. Now, if we take the first, it follows that when there is a flood on one side of the sea, there must be at the same time an ebb somewhere on the other. To this point therefore the inquiry is brought. Now it has been observed by Acosta and others, after careful research, that on the shores of Florida and the opposite shores of Spain and Africa the floods take place at the same times, and the ebbs take place at the same times also; and not that there is an ebb from the shores of Spain and Africa when there is a flood on the shores of Florida. And yet if you look at it more closely, this does not prove the case in favor of the rising and against the progressive motion. For waters may move in progression, and yet rise upon the opposite shores of the same channel at the same time, as when they are thrust together and driven on from some other quarter. For so it is with rivers, which rise and fall on both banks at the same hours. And yet that motion is clearly one of progression, namely, of the waters entering the mouth of the rivers from the sea. It may therefore happen in a like manner that waters coming in a vast mass from the East Indian Ocean are driven together and pushed into the channel of the Atlantic, and on that account flood both sides at once. We must inquire therefore whether there be any other channel in which the water can be retreating and ebbing at that same time; and we have the South Sea, a sea at least as wide, indeed wider and larger than the Atlantic, which is sufficient for the purpose.

At length then, we have come to an instance of the finger-post in this case, and it is this. If we find for certain that when there is a flood on the opposite coasts of Florida and Spain in the Atlantic, there is also a flood on the coasts of Peru and the back of China in the South Sea, then indeed on the authority of this decisive instance we must reject the assertion that the ebb and flow of the sea, which is the thing inquired into, takes place by a progressive motion; for there is no sea or place left in which the retreat or ebbing can be going on at the same time. And this may be most conveniently ascertained by asking the inhabitants of Panama and Lima (where the two oceans, the Atlantic and Pacific, are separated by a small isthmus) whether the ebb and flow of the sea takes place on the opposite sides of the isthmus at the same time; or contrariwise, when it is ebbing on one side it is flowing on the other. Now this decision or rejection appears to be certain, if we take it for granted that the earth is immovable. But if the earth revolves, it is perhaps possible that in consequence of the unequal rotation (in point of speed) of the earth and waters of the sea, the waters are violently driven upwards into a heap, which is the flood, and (when they can bear no more piling) released and let down again, which is the ebb. But on this inquiry should be made separately. Still, even on this hypothesis, our position remains equally fixed, that there must of necessity be an ebb of the sea going on in some parts at the same time that a flood is going on in others.

Again, let the nature in question be the latter of the two motions we have supposed, namely, the rising and sinking motion, if on careful examination we reject the former motion of which I spoke—the progressive. With regard to this nature the road branches into three. For the motion by which the waters rise in the flood and sink in the ebb without any accession of other waters rolling in, must necessarily be brought about in one of these three ways. Either there is an accession of water poured out from the interior of the earth, and again retreating into it; or there is no accession to the mass of water, but the same waters (without increase of quantity) are extended or rarefied so as to occupy a greater space and dimension, and again contract themselves; or there is no increase either of supply or of extension, but the same waters (the same in quantity as in density) are raised by some magnetic force attracting them from above, and by consent therewith, and then fall back again. Let us now dismiss the two former causes of motion and reduce our inquiry to the last; that is to say, let us inquire whether any such raising by consent or magnetic force may happen. Now in the first place it is evident that the waters, as they lie in the trench or hollow of the sea, cannot all be raised at once for want of something to take their place at the bottom; so that even if there were in water any such desire to rise, it would be barred and checked by the cohesion of things, or (as it is commonly called) the abhorrence of a vacuum. It remains that the waters must be raised in one part, and thereby be diminished and retreat in another. Again, it will follow of necessity that the magnetic force, since it cannot act upon the whole, will act with the greatest intensity on the middle, so as to raise up the water in the middle; upon which the rest must follow and fall away from the sides.

Thus at length we come to an instance of the fingerpost on this subject. For if we find that in the ebb of the sea the surface of the water is more arched and round, the waters rising in the middle of the sea and falling away from the sides, that is, the shores; and that in the flood the same surface is more even and level, the waters returning to their former position; then indeed on the authority of this decisive instance the raising by magnetic force may be admitted; otherwise it must be utterly rejected. And this would not be difficult to ascertain by trial in straits with sounding lines, viz., whether during ebbs the sea be not higher or deeper toward the middle than during floods. It is to be observed however that, if this be the case, the waters must (contrary to the common opinion) rise in ebbs and sink in floods, so as to clothe and wash the shores.

Again, let the nature investigated be the spontaneous motion of rotation, and in particular whether the diurnal motion whereby to our eyes the sun and stars rise and set, be a real motion of rotation in the heavenly bodies, or a motion apparent in the heavenly bodies, and real in the earth. We may here take for an instance of the fingerpost the following. If there be found in the ocean any motion from east to west, however weak and languid; if the same motion be found a little quicker in the air, especially within the tropics, where because of the larger circles it is more perceptible; if the same motion be found in the lower comets, but now lively and vigorous; if the same motion be found in planets, but so distributed and graduated that the nearer a planet is to the earth its motion is slower, the further a planet is distant from the earth its motion is quicker, and quickest of all in the starry sphere; then indeed we should receive the diurnal motion as real in the heavens, and deny such motion to the earth. Because it will be manifest that motion from east to west is perfectly cosmical, and by consent of the universe, being most rapid in the highest parts of the heavens, and gradually falling off, and finally stopping and becoming extinct in the immovable—that is, the earth.

Again, let the nature in question be that other motion of rotation so much talked of by philosophers, the resistant and contrary motion to the diurnal, viz., from west to east, which old philosophers attribute to the planets, also to the starry sphere, but Copernicus and his followers to the earth as well. And let us inquire whether any such motion be found in nature, or whether it be not rather a thing invented and supposed for the abbreviation and convenience of calculation, and for the sake of that pretty notion of explaining celestial motions by perfect circles. For this motion in the heavens is by no means proved to be true and real, either by the failing of a planet to return in its diurnal motion to the same point of the starry sphere, or by this, that the poles of the zodiac differ from the poles of the world; to which two things we owe this idea of motion. For the first phenomenon is well accounted for by supposing that the fixed stars outrun the planets and leave them behind; the second, by supposing a motion in spiral lines; so that the inequality of return and the declination to the tropics may rather be modifications of the one diurnal motion than motions contrary or round different poles. And most certain it is, if one may but play the plain man for a moment (dismissing the fancies of astronomers and schoolmen, whose way it is to overrule the senses, often without reason, and to prefer what is obscure), that this motion does actually appear to the sense such as I have described; for I once had a machine made with iron wires to represent it.

The following would be an instance of the fingerpost on this subject. If it be found in any history worthy of credit that there has been any comet, whether high or low, which has not revolved in manifest agreement (however irregular) with the diurnal motion, but has revolved in the opposite direction, then certainly we may set down thus much as established, that there *may be* in nature some such motion. But if nothing of the kind can be found, it must be regarded as questionable, and recourse be had to other instances of the fingerpost about it.

Again, let the nature in question be weight or heaviness. Here the road will branch into two, thus. It must needs be that heavy and weighty bodies either tend of their own nature to the center of the earth, by reason of their proper configuration; or else that they are attracted by the mass and body of earth itself as by the congregation of kindred substances, and move to it by sympathy. If the latter of these be the cause, it follows that the nearer heavy bodies approach to the earth, the more rapid and violent is their motion to it; and that the further they are from the earth, the feebler and more tardy is their motion (as is the case with magnetic attraction); and that this action is confined to certain limits. So that if they were removed to such a distance from the earth that the earth's virtue could not act upon them, they would remain suspended like the earth itself, and not fall at all. With regard to this, then, the following would be an instance of the fingerpost. Take a clock moved by leaden weights, and another moved by the compression of an iron spring. Let them be exactly adjusted, that one go not faster or slower than the other. Then place the clock moving by weights on the top of a very high steeple, keeping the other down below, and observe carefully whether the clock on the steeple goes more slowly than it did on account of the diminished virtue of its weights. Repeat the experiment in the bottom of a mine, sunk to a great depth below the ground; that is, observe whether the clock so placed does not go faster than it did on account of the increased virtue of its weights. If the virtue of the weights is found to be diminished on the steeple and increased in the mine, we may take the attraction of the mass of the earth as the cause of weight.

Again, let the nature investigated be the polarity of the iron needle when touched with the magnet. With regard to this nature the road will branch into two, thus. Either the touch of the magnet of itself invests the iron with polarity to the north and south; or it simply excites and prepares the iron, while the actual motion is communicated by the presence of the earth, as Gilbert thinks, and labors so strenuously to prove. To this point therefore tend the observations which he has collected with great sagacity and industry. One is, that an iron nail which has lain for a long time in a direction between north and south gathers polarity without the touch of the magnet by its long continuance in this position; as if the earth itself, which on account of the distance acts but feebly (the surface or outer crust of the earth being destitute, as he insists, of magnetic power), were yet able by this long continuance to supply the touch of the magnet and excite the iron, and then shape and turn it when excited. Another is, that if iron that has been heated white-hot be, while cooling, laid lengthwise between north and south, it also acquires polarity without the touch of the magnet; as if the parts of the iron, set in motion by ignition and afterwards recovering themselves, were at the very moment of cooling more susceptible and sensitive to the virtue emanating from the earth than at other times, and thus became excited by it. But these things, though well observed, do not quite prove what he asserts.

Now with regard to this question an instance of the fingerpost would be the following. Take a magnetic globe and mark its poles; and set the poles of the globe toward the east and west, not toward the north and south, and let them remain so. Then place at the top an untouched iron needle, and allow it to remain in this position for six or seven days.

The needle while over the magnet (for on this point there is no dispute) will leave the poles of the earth and turn toward the poles of the magnet; and therefore, as long as it remains thus, it points east and west. Now if it be found that the needle, on being removed from the magnet and placed on a pivot, either starts off at once to the north and south, or gradually turns in that direction, then the presence of the earth must be admitted as the cause; but if it either points as before east and west, or loses its polarity, this cause must be regarded as questionable, and further inquiry must be made.

Again, let the nature in question be the corporeal substance of the moon; that is, let us inquire whether it be rare, consisting of flame or air, as most of the old philosophers opined, or dense and solid, as Gilbert and many moderns, with some ancients, maintain. The reasons for the latter opinion rest chiefly on this, that the moon reflects the rays of the sun; nor does light seem to be reflected except by solid bodies. Therefore instances of the fingerpost on this question will (if any) be those which prove that reflection may take place from a rare body, as flame, provided it be of sufficient denseness. Certainly, one cause of twilight, among others, is the reflection of the rays of the sun from the upper part of the air. Likewise we occasionally see rays of the sun in fine evenings reflected from the fringes of dewy clouds with a splendor not inferior to that reflected from the body of the moon, but brighter and more gorgeous; and yet there is no proof that these clouds have coalesced into a dense body of water. Also we observe that the dark air behind a window at night reflects the light of a candle, just as a dense body would. We should also try the experiment of allowing the sun's rays to shine through a hole on some dusky bluish flame. For indeed the open rays of the sun, falling on the duller kinds of flame, appear to deaden them so that they seem more like white smoke than flame. These are what occur to me at present as instances of the fingerpost with reference to this question, and better may perhaps be found. But it should always be observed that reflection from flame is not to be expected, except from a flame of some depth, for otherwise it borders on transparency. This however may be set down as certain—that light on an even body is always either received and transmitted or reflected.

Again, let the nature in question be the motion of projectiles (darts, arrows, balls, etc.) through the air. This motion the schoolmen, as their way is, explain in a very careless manner, thinking it enough to call it a violent motion as distinguished from what they call a natural motion; and to account for the first percussion or impulse by the axiom that two bodies cannot occupy the same place on account of the impenetrability of matter, and not troubling themselves at all how the motion proceeds afterward. But with reference to this inquiry the road branches into two in this way. Either this motion is caused by the air carrying the projected body and collecting behind it, as the stream in the case of a boat, or the wind in that of straws; or it is caused by the parts of the body itself not enduring the impression, but pushing forward in succession to relieve themselves from it. The former of these explanations is adopted by Fracastorius and almost all who have entered into the investigation with any subtlety, and there is no doubt that the air *has* something to do with it. But the other notion is undoubtedly the true one, as is shown by countless experiments. Among others the following would be an instance of the fingerpost on this subject: that a thin iron plate or stiffish iron wire, or even a reed or pen split in half, when pressed into a curve between the finger and thumb, leaps away. For it is obvious that this motion cannot be imputed to the air gathering behind the body, because the source of motion is in the middle of the plate or reed, not in the extremities.

Again, let the nature in question be the rapid and powerful motion of the expansion of gunpowder into flame, by which such vast masses are upheaved, such great weights discharged, as we see in mines and mortars. With respect to this nature the road branches into two in this way. The motion is excited either by the mere desire of the body to expand when set on fire, or partly by that and partly by the desire of the crude spirit in the body, which flies rapidly away from the fire and bursts violently from its embrace as from a prison house. The schoolmen and common opinion deal only with the former desire. For men fancy themselves very fine philosophers when they assert that the flame is endowed by its elementary form with a necessity of occupying a larger space than the body had filled when in the form of powder, and that hence the motion ensues. Meanwhile, they forget to notice that although this be true on the supposition that flame is generated, it is yet possible for the generation of flame to be hindered by a mass of matter sufficient to suppress and choke it; so that the case is not reduced to the necessity they insist on. For that expansion must necessarily take place, and that there must needs follow thereon a discharge or removal of the opposing body, if flame be generated, they rightly judge. But this necessity is altogether avoided if the solid mass suppress the flame before it be generated. And we see that flame, especially in its first generation, is soft and gentle, and requires a hollow space wherein to play and try its strength. Such violence therefore cannot be attributed to flame by itself. But the fact is that the generation of these windy flames, or fiery winds as they may be called, arises from a conflict of two bodies of exactly opposite natures; the one being highly inflammable, which is the nature of sulphur, the other abhorring flame, as the crude spirit in niter. So that there ensues a strange conflict, the sulphur kindling into flame with all its might (for the third body, the willow charcoal, does no more than incorporate and combine the other two), while the spirit of the niter bursts forth with all its might and at the same time dilates itself (as air, water, and all crude bodies do when affected by heat), and by thus flying and bursting out fans meanwhile the flame of the sulphur on all sides as with hidden bellows.

On this subject we may have instances of the fingerpost of two kinds. The first, of those bodies which are most highly inflammable, as sulphur, camphor, naphtha and the like, with their compounds, which catch fire more quickly and easily than gunpowder if not impeded (from which it appears that the desire of bursting into flame does not produce by itself that stupendous effect); the other, of those bodies which shun and abhor flame, as all salts. For we find that if salts are thrown into the fire their aqueous spirit bursts out with a crackling noise before flame is caught; which is the case also, though in a milder degree, with the stiffer kinds of leaves, the aqueous part escaping before the oily catches fire. But this is best seen in quicksilver, which is not inaptly called mineral water. For quicksilver, without bursting into flame, by mere eruption and expansion almost equals the force of gunpowder, and is also said, when mixed with gunpowder, to increase its strength.

Again, let the nature in question be the transitory nature of flame and its instantaneous extinction. For the nature of flame appears to have no fixed consistency here with us, to be every moment generated and every moment extinguished; for it is clear that in flames which continue and last, the continuance we see is not of the same individual flame, but is caused by a succession of new flame regularly generated. Nor does the flame remain numerically identical, as is easily seen from this, that if the food or fuel of flame be taken away, the flame instantly goes out. With reference to this nature the roads branch into two, thus: the instantaneous nature proceeds either from a cessation of the cause which at first produced the flame, as in light, sound, and the motion called "violent"; or from this, that the flame, though able by its own nature to remain with us, suffers violence and is destroyed by the contrary natures that surround it.

On this subject therefore we may take the following as an instance of the fingerpost. We see in large fires how high the flames ascend, for the broader the base of the flame, the higher is its vertex. Thus extinction appears to commence at the sides, where the flame is compressed and troubled by the air. But the heart of the flame, which is not touched by the air but surrounded by other flame on all sides, remains numerically identical; nor is it extinguished until gradually compressed by the surrounding air. Thus all flame is in the form of a pyramid, being broader at the base where the fuel is, but sharp at the vertex, where the air is antagonistic and fuel is wanting. But smoke is narrow at the base and grows broader as it ascends, like an inverted pyramid; the reason being that the air admits smoke and compresses flame. For let no one dream that lighted flame is air, when in fact they are substances quite heterogeneous.

But we may have an instance of the fingerpost more nicely adapted to this purpose, if the thing can be made manifest with bicolored lights. Fix a lighted wax taper in a small metal stand; place the stand in the middle of a bowl, and pour round it spirit of wine, but not enough to reach the top of the stand. Then set fire to the spirit of wine. The spirit of wine will yield a bluish, the taper a yellow flame. Observe therefore whether the flame of the taper (which is easily distinguished by its color from the flame of the spirit of wine, since flames do not mix at once, as liquids do) remains in a conical or rather tends to a globular form, now that there is nothing to destroy or compress it. If the latter is found to be the case, it may be set down as certain that flame remains numerically identical as long as it is enclosed within other flame and feels not the antagonistic action of the air.

Let this suffice for instances of the fingerpost. I have dwelt on them at some length to the end that men may gradually learn and accustom themselves to judge of nature by instances of the fingerpost and experiments of light, and not by probable reasonings.

XXXVII

Among Prerogative Instances I will put in the fifteenth place *Instances of Divorce,* which indicate the separation of natures of most familiar occurrence. They differ from the instances subjoined to the instances of companionship, in that the latter indicate the separation of a nature from some concrete substance with which it is ordinarily in conjunction, while these instances indicate the separation of one nature from another. They differ from instances of the fingerpost, in that they determine nothing, but simply notify the separability of one nature from another. Their use is to detect false forms and to dissipate slight theories suggested by what lies on the surface, and so serve as ballast to the understanding.

For example, let the natures investigated be those four natures which Telesius accounts as messmates and chamber fellows, namely: heat, brightness, rarity, mobility or promptness to motion. We find, however, many instances of divorce between them. For air is rare and mobile, not hot or bright; the moon is bright without heat; boiling water is hot without light; the motion of an iron needle on a pivot is quick and nimble, and yet the body is cold, dense, and opaque; and there are many more of the kind.

Again, let the natures investigated be corporeal nature and natural action. For it seems that natural action is not found except as subsisting in some body. Yet in this case also we shall perhaps be able to find some instance of divorce; such, for example, as magnetic action, by which iron is drawn to the magnet, heavy bodies to the globe of the earth. There may also be added some other operations performed at a distance. For such action takes place both in time, occupying moments not a mere instant of time, and in space, passing through degrees and distances. There is therefore some moment of time, and some distance of space, in which the virtue or action remains suspended between the two bodies which produce the motion. The question therefore is brought to this: whether the bodies which are the limits of the motion dispose or alter the intermediate bodies, so that by a succession of actual contacts the virtue passes from limit to limit, meanwhile subsisting in the intermediate body; or whether there is no such thing, but only the bodies, the virtue, and the distances. In rays of light, indeed, and sounds, and heat, and certain other things acting at a distance, it is probable that the intermediate bodies are disposed and altered, the more so because they require a medium qualified for carrying on the operation. But that magnetic or attractive virtue admits of media without distinction, nor is the virtue impeded in any kind of medium. And if the virtue or action has nothing to do with the intermediate body, it follows that there is a natural virtue or action subsisting for a certain time and in a certain space without a body, since it neither subsists in the limiting nor in the intermediate bodies. And therefore magnetic action may be an instance of divorce between corporeal nature and natural action. To which may be appended as a corollary or advantage not to be omitted that here is a proof furnished by merely human philosophy of the existence of essences and substances separate from matter and incorporeal. For allow that natural virtue and action, emanating from a body, can exist for a certain time and in a certain space altogether without a body, and you are not far from allowing that it can also emanate originally from an incorporeal substance. For corporeal nature appears to be no less requisite for sustaining and conveying natural action than for exciting or generating it.

XXXVIII

Now follow five classes of instances which under one general name I call *Instances of the Lamp,* or *of First Information.* They are those which aid the senses. For since all interpretation of nature commences with the senses and leads from the perceptions of the senses by a straight, regular, and guarded path to the perceptions of the understanding, which are true notions and axioms, it follows of necessity that the more copious and exact the representations of the senses, the more easily and prosperously will everything proceed.

Of these five instances of the lamp, the first strengthen, enlarge, and rectify the immediate actions of the senses; the second make manifest things which are not directly perceptible by means of others which are; the third indicate the continued processes or series of those things and motions which are for the most part unobserved except in their end or periods; the fourth provide the sense with some substitute when it utterly fails; the fifth excite the attention and notice of the sense, and at the same time set bounds to the subtlety of things. Of these I shall now speak in their order.

XXXIX

Among Prerogative Instances I will put in the sixteenth place *Instances of the Door or Gate,* this being the name I give to instances which aid the immediate actions of the senses. Now of all the senses it is manifest that sight has the chief office in giving information. This is the sense, therefore, for which we must chiefly endeavor to procure aid. Now the aids to sight are of three kinds: it may be enabled to perceive objects that are not visible; to perceive them further off; and to perceive them more exactly and distinctly.

Of the first kind (not to speak of spectacles and the like, which serve only to correct or relieve the infirmity of a defective vision, and therefore give no more information) are those recently invented glasses which disclose the latent and invisible minutiae of bodies and their hidden configurations and motions by greatly increasing their apparent size; instruments by the aid of which the exact shape and outline of body in a flea, a fly, a worm, and also colors and motions before unseen, are not without astonishment discerned. It is also said that a straight line drawn with a pen or pencil is seen through such glasses to be very uneven and crooked, the fact being that neither the motion of the hand, though aided by a ruler, nor the impression of the ink or color, is really even, although the unevenness is so minute that it cannot be detected without such glasses. And here (as is usual in things new and wonderful) a kind of superstitious observation has been added, viz., that glasses of this sort do honor to the works of nature but dishonor to the works of art. The truth however is only this, that natural textures are far more subtle than artificial. For the microscope, the instrument I am speaking of, is only available for minute objects. So that if Democritus had seen one, he would perhaps have leaped for joy, thinking a way was now discovered of discerning the atom, which he had declared to be altogether invisible. The incompetency however of such glasses, except for minutiae alone, and even for them when existing in a body of considerable size, destroys the use of the invention. For if it could be extended to larger bodies, or to the minutiae of larger bodies, so that the texture of a linen cloth could be seen like network, and thus the latent minutiae and inequalities of gems, liquors, urine, blood, wounds, etc., could be distinguished, great advantages might doubtless be derived from the discovery.

Of the second kind are those other glasses discovered by the memorable efforts of Galileo, by the aid of which, as by boats or vessels, a nearer intercourse with the heavenly bodies can be opened and carried on. For these show us that the Milky Way is a group or cluster of small stars entirely separate and distinct, of which fact there was but a bare suspicion among the ancients. They seem also to point out that the spaces of the planetary orbits, as they are called, are not altogether destitute of other stars, but that the heaven begins to be marked with stars before we come to the starry sphere itself, although with stars too small to be seen without these glasses. With this instrument we can descry those small stars wheeling as in a dance round the planet Jupiter, whence it may be conjectured that there are several centers of motion among the stars. With this the inequalities of light and shade in the moon are more distinctly seen and placed, so that a sort of selenography can be made. With this we descry spots on the sun, and similar phenomena—all indeed noble discoveries, so far as we may safely trust to demonstrations of this kind, which I regard with suspicion chiefly because the experiment stops with these few discoveries, and many other things equally worthy of investigation are not discovered by the same means.

Of the third kind are measuring rods, astrolabes, and the like, which do not enlarge the sense of sight, but rectify and direct it. And if there are other instances which aid the remaining senses in their immediate and individual actions, and yet are of a kind which add nothing to the information already possessed; they are not to the present purpose, and therefore I have omitted to mention them.

XL

Among Prerogative Instances I will put in the seventeenth place *Summoning Instances,* borrowing the name from the courts of law, because they summon objects to appear which have not appeared before. I also call them *Evoking Instances.* They are those which reduce the nonsensible to the sensible, that is, make manifest things not directly perceptible by means of others which are.

An object escapes the senses either on account of its distance; or on account of the interposition of intermediate bodies; or because it is not fitted for making an impression on the sense; or because it is not sufficient in quantity to strike the sense; or because there is not time enough for it to act on the sense; or because the impression of the object is such as the sense cannot bear; or because the sense has been previously filled and occupied by another object, so that there is not room for a new motion. These cases have reference principally to the sight, and secondarily to the touch. For these two senses give information at large and concerning objects in general, whereas the other three give hardly any information but what is immediate and relates to their proper objects.

In the first kind, where an object is imperceptible by reason of its distance, there is no way of manifesting it to the sense but by joining to it or substituting for it some other object which may challenge and strike the sense from a greater distance—as in communication by beacons, bells, and the like.

In the second kind, this reduction or secondary manifestation is effected when objects that are concealed by the interposition of bodies within which they are enclosed and cannot conveniently be opened out are made manifest to the sense by means of those parts of them which lie on the surface, or make their way from the interior. Thus the condition of the human body is known by the state of the pulse, urine, and the like.

In the third and fourth kind, reductions are applicable to a great many things, and in the investigations of nature should be sought for on all sides. For example, it is obvious that air and spirit, and like bodies, which in their entire substance are rare and subtle, can neither be seen nor touched. Therefore, in the investigation of bodies of this kind it is altogether necessary to resort to reductions.

Thus let the nature in question be the action and motion of the spirit enclosed in tangible bodies. For everything tangible that we are acquainted with contains an invisible and intangible spirit which it wraps and clothes as with a garment. Hence that three-fold source, so potent and wonderful, of the process of the spirit in a tangible body. For the spirit in a tangible substance, if discharged, contracts bodies and dries them up; if detained, softens and melts them; if neither wholly discharged nor wholly detained, gives them shape, produces limbs, assimilates, digests, ejects, organizes, and the like. And all these processes are made manifest to the sense by conspicuous effects.

For in every tangible inanimate body the enclosed spirit first multiplies itself and, as it were, feeds upon those tangible parts which are best disposed and prepared for that purpose and so digests and elaborates and turns them into spirit; and then they escape together. Now this elaboration and multiplication of the spirit is made manifest to the sense by diminution of weight. For in all desiccation there is some decrease of quantity, not only of the quantity of spirit previously existing in the body, but also of the body itself, which was before tangible and is newly changed. For spirit is without weight. Now the discharge or emission of the spirit is made manifest to the sense in the rust of metals and other similar putrefactions which stop short before they come to the rudiments of life; for these belong to the third kind of process. For in compact bodies the spirit finds no pores or passages through which to

escape and is therefore compelled to push and drive before it
the tangible parts themselves, so that they go out along with
it; whence proceed rust and the like. On the other hand the
contraction of the tangible parts after some of the spirit is discharged (upon which desiccation ensues), is made manifest to
the sense not only by the increased hardness of the body, but
much more by the rents, contractions, wrinklings, and shrivel-
ings in the body which thereupon take place. For the parts of
wood split asunder and are contracted; skins shrivel; and not
only that, but if the spirit is suddenly discharged by the heat
of fire, they hasten so fast to contraction as to curl and roll
themselves up.

On the contrary, where the spirit is detained and yet expanded and excited by heat or something analogous thereto
(as happens in the more solid or tenacious bodies), then are
bodies softened, as white hot iron; or they become fluid, as
metals; or liquid, as gums, wax, and the like. Thus the contrary operations of heat, which hardens some substances and
melts others, are easily reconciled, since in the former the
spirit is discharged, in the latter it is excited and detained;
whereof the melting is the proper action of the heat and spirit,
the hardening is the action of the tangible parts only on occasion of the discharge of the spirit.

But when the spirit is neither wholly detained nor wholly
discharged, but only makes trials and experiments within its
prison house, and meets with tangible parts that are obedient
and ready to follow, so that wheresoever the spirit leads they
go along with it, then ensues the forming of an organic body
and the development of organic parts, and all the other vital
actions as well in vegetable as in animal substances. And these
operations are made manifest to the sense chiefly by careful
observation of the first beginnings and rudiments or essays of
life in animalculae generated from putrefaction, as in ants'
eggs, worms, flies, frogs after rain, etc. There is required, however, for the production of life both mildness in the heat and
pliancy in the substance, that the spirit may neither be so hur-

ried as to break out, nor be confined by the obstinacy of the parts, but may rather be able to mold and model them like wax.

Again, that most noble distinction of spírit which has so many applications (viz., spirit cut off; spirit simply branching; spirit at once branching and cellulate—of which the first is the spirit of all inanimate substances, the second of vegetables, the third of animals), is brought as it were before the eyes by several instances of this kind of reduction.

In like manner it appears that the more subtle textures and configurations of things (though the entire body be visible or tangible) are perceptible neither to the sight nor touch. And therefore in these also, our information comes by way of reduction. Now the most radical and primary difference between configurations is drawn from the abundance or scantiness of the matter occupying the same space or dimensions. For all other configurations (which have reference to the dissimilarity of the parts contained in the same body, and to their collocation and position) are but secondary in comparison with the former.

Thus let the nature in question be the expansion or coition of matter in bodies compared one with another, viz., how much matter occupies how much space in each. For there is nothing more true in nature than the twin propositions that "nothing is produced from nothing," and "nothing is reduced to nothing," but that the absolute quantum or sum total of matter remains unchanged, without increase or diminution. Nor is it less true that of that quantum of matter more or less is contained under the same space or dimensions according to the diversity of bodies; as in water more, in air less. So that to assert that a given volume of water can be changed into an equal volume of air is as much as to say that something can be reduced to nothing; as on the other hand to maintain that a given volume of air can be turned into an equal volume of water is the same as to say that something can be produced out of nothing. And it is from this abundance and scantiness of matter that the abstract notions of dense and rare, though variously and promiscuously used, are, properly speaking, derived. We must also take for granted a third proposition which is also sufficiently certain, viz., that this greater or less quantity of matter in this or that body is capable of being reduced by comparison to calculation and to exact or nearly exact proportions. Thus one would be justified in asserting that in any given volume of gold there is such an accumulation of matter, that spirit of wine, to make up an equal quantity of matter, would require twenty-one times the space occupied by the gold.

Now the accumulation of matter and its proportions are made manifest to the sense by means of weight. For the weight answers to the quantity of matter in the parts of a tangible body, whereas spirit and the quantum of matter which it contains cannot be computed by weight, for it rather diminishes the weight than increases it. But I have drawn up a very accurate table on this subject, in which I have noted down the weights and volumes of all the metals, the principal stones, woods, liquors, oils, and many other bodies, natural as well as artificial—a thing of great use in many ways, as well for light of information as for direction in practice, and one that discloses many things quite beyond expectation. Not the least important of which is this—it shows that all the variety in tangible bodies known to us (such bodies I mean as are tolerably compact and not quite spongy and hollow, and chiefly filled with air) does not exceed the limit of the ratio of 1 to 21—so limited is nature, or at any rate that part of it with which we have principally to do.

I have also thought it worth while to try whether the proportions can be calculated which intangible or pneumatic bodies bear to bodies tangible. This I attempted by the following contrivance. I took a glass phial, capable of holding about an ounce, using a small vessel that less heat might be required to produce evaporation. This phial I filled with spirit of wine almost to the neck, selecting spirit of wine, because I found by the former table that of all tangible bodies (which are well united and not hollow) this is the rarest and contains the least quantity of matter in a given space. After that, I noted exactly the weight of the spirit and phial together. I then took a bladder capable of holding about a quart from which I squeezed out, as well as I could, all the air, until the two sides of the bladder met. The bladder I had previously rubbed over gently with oil, to make it closer, and having thus stopped up the pores, if there were any, I inserted the mouth of the phial within the mouth of the bladder, and tied the latter tightly round the former with a thread smeared with wax in order that it might stick more closely and tie more firmly. After this I set the phial on a chafing dish of hot coals. Presently the steam or breath of the spirit of wine, which was dilated and rendered pneumatic by the heat, began gradually to expand the bladder and swelled it out on all sides like a sail. When this took place, I immediately took the glass off the fire, placing it on a carpet that it might not crack with the cold, at the same time making a hole in the bladder lest the steam should turn liquid again on the cessation of the heat and so disturb the calculations. I then removed the bladder, and weighing the spirit of wine which remained, computed how much had been converted into steam or air. Then, comparing the space which the body had occupied while it was spirit of wine in the phial with the space which it afterward occupied when it had become pneumatic in the bladder, I computed the results, which showed clearly that the body had acquired by the change a degree of expansion a hundred times greater than it had had before.

Again, let the nature in question be heat or cold, in a degree too weak to be perceptible to the sense. These are made manifest to the sense by a calendar glass such as I have described above. For the heat and cold are not themselves perceptible to the touch, but the heat expands the air, and the cold contracts it. Nor again is this expansion and contraction of the air perceptible to the sight, but the expansion of the air depresses the water, the contraction raises it, and so at last is made manifest to the sight; not before, nor otherwise.

Again, let the nature in question be the mixture of bodies, viz., what they contain of water, oil, spirit, ash, salt, and the like; or (to take a particular instance) what quantity of butter, curd, whey, etc., is contained in milk. These mixtures, so far as relates to tangible elements, are made manifest to the sense by artificial and skillful separations. But the nature of the spirit in them, though not immediately perceived, is yet discovered by the different motions and efforts of the tangible bodies in the very act and process of their separation and also by the acridities and corrosions, and by the different colors, smells, and tastes of the same bodies after separation. And in this department men have labored hard, it is true, with distillations and artificial separations, but not with much better success than in the other experiments which have been hitherto in use. For they have but groped in the dark and gone by blind ways and with efforts painstaking rather than intelligent, and (what is worst of all), without attempting to imitate or emulate nature, but rather destroying by the use of violent heats and overstrong powers all that more subtle configuration in which the occult virtues and sympathies of things chiefly reside. Nor do they remember or observe, while making such separations, the circumstances which I have elsewhere pointed out, namely, that when bodies are tormented by fire or other means, many qualities are communicated by the fire itself and by the bodies employed to effect the separation which did not exist previously in the compound; whence strange fallacies have arisen. For it must not be supposed that all the vapor which is discharged from water by the action of fire was formerly vapor or air in the body of the water, the fact being that the greatest part of it was created by the expansion of the water from the heat of the fire.

So in general, all the nice tests of bodies whether natural or artificial by which the genuine are distinguished from the adulterated, the better from the viler sort, should be referred to this division; for they make manifest to the sense things not directly perceptible by means of those which are. They should therefore be sought and collected from all quarters with diligent care.

With regard to the fifth way in which objects escape the

sense, it is obvious that the action of sense takes place in motion, and that motion takes place in time. If therefore the motion of any body be either so slow or so quick that it bears no proportion to the moments which the sense takes to act in, the object is not perceived at all, as in the motion of the hand of a clock and again in the motion of a musket ball. Now motion which is too slow to be perceived is easily and usually made manifest to the sense by means of aggregates of motion. Motion which is too quick has not hitherto been competently measured, and yet the investigation of nature requires that this be done in some cases.

In the sixth kind, where the sense is hindered by the too great power of the object, the reduction may be effected either by removing the object to a greater distance from the sense; or by deadening its effects by the interposition of a medium which will weaken without annihilating the object; or by admitting and receiving the reflection of the object where the direct impression is too powerful, as that of the sun, for instance, in a basin of water.

The seventh cause, where the sense is so charged with one object that it has no room for the admission of another, is almost wholly confined to the sense of smell and has little to do with the matter in hand. So much then for the reduction of the nonsensible to the sensible—or the modes of making manifest to the sense things not directly perceptible by means of others which are.

Sometimes, however, the reduction is made not to the sense of a man, but of some other animal whose sense in some cases is keener than man's; as of certain scents to the sense of a dog; of the light which is latent in air when not illumined from without to the sense of a cat, owl, and similar animals which see in the dark. For Telesius has justly observed that there is in the air itself a certain original light, though faint and weak, and hardly of any use to the eyes of men and most animals; inasmuch as animals to whose sense this light is adapted see in the dark, which it is hardly to be believed they do either without light, or by a light within.

Observe also that at present I am dealing with the deficiencies of the senses and their remedies. The deceptions of the senses must be referred to the particular inquiries concerning sense and the objects of sense, excepting only that grand deception of the senses, in that they draw the lines of nature with reference to man and not with reference to the universe; and this is not to be corrected except by reason and universal philosophy.

XLI

Among Prerogative Instances I will put in the eighteenth place *Instances of the Road,* which I also call *Traveling Instances* and *Articulate Instances.* They are those which point out the motions of nature in their gradual progress. This class of instances escapes the observation rather than the sense. For it is strange how careless men are in this matter; for they study nature only by fits and at intervals, and when bodies are finished and completed, not while she is at work upon them. Yet if anyone were desirous of examining and studying the contrivances and industry of an artificer, he would not be content with beholding merely the rude materials of the art and then the completed works, but would rather wish to be present while the artificer was at his labors and carrying his work on. And a like course should be taken with the investigation of nature. For instance, if we are inquiring into the vegetation of plants, we must begin from the very sowing of the seed, and observe (as we may easily do, by taking out day after day the seeds that have lain in the ground two days, three days, four days, and so on, and carefully examining them) how and when the seed begins to puff and swell and to be, as it were, filled with spirit; secondly, how it begins to burst the skin and put forth fibers, at the same time raising itself slightly upwards, unless the ground be very stiff; also, how it puts forth its fibers, some for the root downwards and some for the stem upwards, and sometimes also creeping sideways if it there finds the ground more open and yielding; and so with many other things of the kind. In the same way we should examine the hatching of eggs, in which we might easily observe the whole process of vivification and organization, and see what parts proceed from the yolk and what from the white of the egg, and so forth. A similar course should be taken with animals generated from putrefaction. For to prosecute such inquiries concerning perfect animals by cutting out the fetus from the womb would be too inhuman, except when opportunities are afforded by abortions, the chase, and the like. There should therefore be set a sort of night watch over nature, as showing herself better by night than by day. For these may be regarded as night studies by reason of the smallness of our candle and its continual burning.

The same too should be attempted with inanimate substances, as I have done myself in investigating the expansion of liquids by fire. For there is one mode of expansion in water, another in wine, another in vinegar, another in verjuice, and quite another in milk and oil; as was easily to be seen by boiling them over a slow fire and in a glass vessel in which everything may be clearly distinguished. These matters, however, I touch but briefly, meaning to treat of them more fully and exactly when I come to the discovery of the *Latent Process* of things. For it should all along be borne in mind that in this place I am not handling the things themselves, but only giving examples.

XLII

Among Prerogative Instances I will put in the nineteenth place *Supplementary* or *Substitutive Instances*, which I also call *Instances of Refuge*. They are those which supply information when the senses entirely fail us, and therefore we fly to them when appropriate instances are not to be had. Now substitution is made in two ways: either by gradual approximation or by analogy. To take an example: There is no medium known by the interposition of which the operation of the magnet in drawing iron is entirely prevented. Gold placed between does not stop it, nor silver, nor stone, nor glass, wood, water, oil, cloth or fibrous substances, air, flame, etc. But yet by nice tests some medium may possibly be found to deaden its virtue more than any other; comparatively, that is, and in some degree. Thus it may be that the magnet would not attract iron as well through a mass of gold as through an equal space of air, or through ignited silver as well as through cold; and so in other cases. For I have not made the trial myself in these cases. It is enough to propose such experiments by way of example. Again, there is no body we are acquainted with which does not contract heat on being brought near the fire. And yet air contracts heat much more quickly than stone. Such is the substitution which is made by gradual approximation.

Substitution by analogy is doubtless useful, but is less certain, and should therefore be applied with some judgment. It is employed when things not directly perceptible are brought within reach of the sense, not by perceptible operations of the imperceptible body itself, but by observation of some cognate body which is perceptible. For example, suppose we are inquiring into the mixture of spirits, which are invisible bodies. There seems to be a certain affinity between bodies and the matter that feeds or nourishes them. Now the food of flame seems to be oil and fat substances; of air, water and watery substances; for flame multiplies itself over exhalations of oil, air over the vapor of water. We should therefore look to the mixture of water and oil, which manifests itself to the sense, since the mixture of air and flame escapes the sense. Now oil and water, which are mingled together very imperfectly by composition or agitation, are in herbs and blood and the parts of animals very subtly and finely mingled. It is possible, therefore, that something similar may be the case with the mixture of flame and air in pneumatic bodies, which, though not readily mingling by simple commixture, yet seem to be mingled together in the spirits of plants and animals, especially as all animate spirit feeds on moist substances of both kinds, watery and fat, as its proper food.

Again, if the inquiry be not into the more perfect mixtures of pneumatic bodies but simply into their composition, that is, whether they be readily incorporated together; or whether there be not rather, for example, certain winds and exhalations or other pneumatic bodies which do not mix with common air, but remain suspended and floating therein in globules and drops and are rather broken and crushed by the air than admitted into or incorporated with it—this is a thing which cannot be made manifest to the senses in common air and other pneumatic bodies, by reason of their subtlety. Yet how far the thing may take place we may conceive, by way of image or representation, from what takes place in such liquids as quicksilver, oil, or water, and likewise from the breaking up of air when it is dispersed in water and rises in little bubbles; and again in the thicker kinds of smoke; and lastly, in dust raised and floating in the air; in all of which cases no incorporation takes place. Now the representation I have described is not a bad one for the matter in question, provided that diligent inquiry has been first made whether there can be such a heterogeneity in pneumatic bodies as we find there is in liquids. For if there can, then these images by analogy may not inconveniently be substituted.

But with regard to these supplementary instances, although I stated that information was to be derived from them in the absence of instances proper, as a last resource, yet I wish it to be understood that they are also of great use even when proper instances are at hand—for the purpose, I mean, of corroborating the information which the others supply. But I shall treat of them more fully when I come in due course to speak of the *Supports of Induction.*

XLIII

Among Prerogative Instances I will put in the twentieth place *Dissecting Instances,* which I also call *Awakening Instances,* but for a different reason. I call them awakening, because they awaken the understanding; dissecting, because they dissect nature. For which reason also I sometimes call them *Democritean*. They are those which remind the understanding of the wonderful and exquisite subtlety of nature, so as to stir it up and awaken it to attention and observation and due investigation. Such, for example, as these following: that a little drop of ink spreads to so many letters or lines; that silver gilt stretches to such a length of gilt wire; that a tiny worm, such as we find in the skin, possesses in itself both spirit and a varied organization; that a little saffron tinges a whole hogshead of water; that a little civet or musk scents a much larger volume of air; that a little incense raises such a cloud of smoke; that such exquisite differences of sounds, as articulate words, are carried in every direction through the air, and pierce even, though considerably weakened, through the holes and pores of wood and water, and are moreover echoed back, and that too with such distinctness and velocity; that light and color pass through the solid substances of glass and water so speedily, and in so wide an extent, and with such copious and exquisite variety of images, and are also refracted and reflected; that the magnet acts through bodies of all sorts, even the most compact; and yet (which is more strange) that in all these, passing as they do through an indifferent medium (such as the air is), the action of one does not much interfere with the action of another. That is to say, that at the same time there are carried through spaces of air so many images of visible objects, so many impressions of articulate sound, so many distinct odors, as of a violet, rose, etc.; moreover, heat and cold and magnetic influences—all (I say) at once without impeding one another, just as if they had their own roads and passages set apart, and none ever struck or ran against other.

To these dissecting instances it is useful however to subjoin instances which I call limits of dissection, as that in the cases above mentioned, though one action does not disturb or impede another action of a different kind, yet one action does overpower and extinguish another action of the same kind; as the light of the sun extinguishes that of a glowworm; the report of a cannon drowns the voice; a strong scent overpowers a more delicate one; an intense heat a milder one; a plate of iron interposed between a magnet and another piece of iron destroys the action of the magnet. But this subject also will find its proper place among the supports of induction.

XLIV

So much for instances which aid the senses, instances which are chiefly useful for the informative part of our subject. For information commences with the senses. But the whole business terminates in works, and as the former is the beginning, so the latter is the end of the matter. I will proceed therefore with the instances which are pre-eminently useful for the operative part. They are of two kinds, and seven in number, though I call them all by the general name of *Practical Instances*. In the operative part there are two defects and two corresponding prerogatives of instances. For operation either fails us or it overtasks us. The chief cause of failure in operation (especially after natures have been diligently investigated) is the ill determination and measurement of the forces and actions of bodies. Now the forces and actions of bodies are circumscribed and measured, either by distances of space, or by moments of time, or by concentration of quantity, or. by predominance of virtue. And unless these four things have been well and carefully weighed we shall have sciences fair perhaps in theory, but in practice inefficient. The four instances which are useful in this point of view I class under one head as *Mathematical Instances* and *Instances of Measurement*.

Operation comes to overtask us, either through the admixture of useless matters, or through the multiplicity of instruments, or through the bulk of the material and of the bodies that may happen to be required for any particular work. Those instances therefore ought to be valued which either direct practice to the objects most useful to mankind; or which save instruments; or which spare material and provision. The three instances which serve us here I class together as *Propi-*

tious or *Benevolent Instances.* These seven instances I will now discuss separately, and with them conclude that division of my subject which relates to the Prerogative or Rank of Instances.

XLV

Among Prerogative Instances I will put in the twenty-first place *Instances of the Rod or Rule,* which I also call *Instances of Range* or *of Limitation.* For the powers and motions of things act and take effect at distances not indefinite or accidental, but finite and fixed; so that to ascertain and observe these distances in the investigation of the several natures is of the greatest advantage to practice, not only to prevent its failure but also to extend and increase its power. For we are sometimes enabled to extend the range of powers and, as it were, to diminish distances, as for instance by the use of telescopes.

Most of these powers act and take effect only by manifest contact, as in the impact of two bodies, where the one does not move the other from its place unless they touch each other. Also medicines that are applied externally, as ointments or plasters, do not exert their virtues without touching the body. Finally, the objects of the taste and touch do not strike those senses unless they be contiguous to the organs.

There are also powers which act at a distance, though a very small one; and of these only a few have been hitherto observed, albeit there are many more than men suspect; as (to take common examples) when amber or jet attracts straws; bubbles dissolve bubbles on being brought together; certain purgative medicines draw humors downward, and the like. So, too, the magnetic power by which iron and a magnet, or two magnets, are made to meet, operates within a fixed but narrow sphere of action; but if there be any magnetic virtue flowing from the earth (a little below the surface), and acting on a steel needle in respect of its polarity, the action operates at a great distance.

Again, if there be any magnetic power which operates by consent between the globe of the earth and heavy bodies, or between the globe of the moon and the waters of the sea (as seems highly probable in the semimenstrual ebbs and floods), or between the starry sphere and the planets whereby the latter are attracted to their apogees, all these must operate at very great distances. There are found also certain materials which catch fire a long way off, as we are told the naphtha of Babylon does. Heat also insinuates itself at great distances, as also does cold; insomuch that by the inhabitants of Canada the masses of ice that break loose and float about the northern ocean and are borne through the Atlantic toward that coast are perceived at a great distance by the cold they give out. Perfumes also (though in these there appears to be always a certain corporeal discharge) act at remarkable distances, as those find who sail along the coasts of Florida or some parts of Spain, where there are whole woods of lemon and orange and like odoriferous trees, or thickets of rosemary, marjoram, and the like. Lastly, the radiations of light and impressions of sound operate at vast distances.

But whether the distances at which these powers act be great or small, it is certain that they are all finite and fixed in the nature of things, so that there is a certain limit never exceeded, and a limit which depends either on the mass or quantity of matter in the bodies acted on; or on the strength or weakness of the powers acting; or on the helps or hindrances presented by the media in which they act—all which things should be observed and brought to computation. Moreover, the measurements of violent motions (as they are called), as of projectiles, guns, wheels, and the like, since these also have manifestly their fixed limits, should be observed and computed.

There are found also certain motions and virtues of a contrary nature to those which operate by contact and not at a distance, namely, those which operate at a distance and not by contact; and again those which operate more feebly at a lesser

distance, and more powerfully at a greater. The act of sight
for instance is not well performed in contact but requires a
medium and a distance. Yet I remember being assured by a
person of veracity that he himself under an operation for
the cataract, when a small silver needle was inserted within
the first coat of the eye in order to remove the pellicle of the
cataract and push it into a corner, saw most distinctly the
needle passing over the very pupil. But though this may be
true, it is manifest that large bodies are not well or distinctly
seen except at the vertex of a cone, the rays from the object
converging at a certain distance from it. Moreover, old people
see objects better at a little distance than if quite close. In
projectiles, too, it is certain that the impact is not so violent
at too small a distance as it is a little further off. These, therefore, and like things should be observed in the measurements
of motions with regard to distances.

There is also another kind of local measurement of motions
which must not be omitted. This has to do with motions not
progressive, but spherical, that is, with the expansion of
bodies into a greater sphere or their contraction into a less.
For among our measurements of motions we must inquire
what degree of compression or extension bodies (according to
their nature) easily and freely endure, and at what point they
begin to resist, till at last they will bear no more. Thus, when
a blown bladder is compressed, it allows a certain compression of the air, but if the compression be increased the air
does not endure it and the bladder bursts.

But this same thing I have tested more accurately by a
subtle experiment. I took a small bell of metal, light and
thin, such as is used for holding salt, and plunged it into a
basin of water so that it carried down with it the air contained in its cavity to the bottom of the basin, where I had
previously placed a small globe, on which the bell was to
light. I found then that if the globe was small enough in proportion to the cavity, the air contracted itself into a less
space and was simply squeezed together, not squeezed out. But

if it was too large for the air to yield freely, then the air, impatient of greater pressure, raised the bell on one side and rose to the surface in bubbles.

Again, to test the extension as well as compression of which air was susceptible, I had recourse to the following device. I took a glass egg with a small hole at one end of it, and, having drawn out the air through the hole by violent suction, I immediately stopped up the hole with my finger and plunged the egg into water, and then took away my finger. The air, having been extended by the suction and dilated beyond its natural dimensions, and therefore struggling to contract itself again (so that if the egg had not been plunged into the water it would have drawn in air with a hissing sound), now drew in water in sufficient quantities to allow the air to recover its old sphere or dimension.

Now it is certain that the rarer bodies (such as air) allow a considerable degree of contraction, as has been stated, but that tangible bodies (such as water) suffer compression with much greater difficulty and to a lesser extent. How far they do suffer it I have investigated in the following experiment. I had a hollow globe of lead made, capable of holding about two pints, and sufficiently thick to bear considerable force. Having made a hole in it, I filled it with water and then stopped up the hole with melted lead, so that the globe became quite solid. I then flattened two opposite sides of the globe with a heavy hammer, by which the water was necessarily contracted into less space, a sphere being the figure of largest capacity. And when the hammering had no more effect in making the water shrink, I made use of a mill or press, till the water, impatient of further pressure, exuded through the solid lead like a fine dew. I then computed the space lost by the compression and concluded that this was the extent of compression which the water had suffered, but only when constrained by great violence.

But the compression or extension endured by more solid, dry, or more compact bodies, such as wood, stones and metals, is still less than this, and scarcely perceptible. For they free themselves either by breaking, or by moving forward, or by other efforts, as is apparent in the bending of wood or metal, in clocks moving by springs, in projectiles, hammerings, and numberless other motions. And all these things with their measures should in the investigation of nature be explored and set down, either in their certitude, or by estimate, or by comparison, as the case will admit.

XLVI

Among Prerogative Instances I will put in the twenty-second place *Instances of the Course,* which I also call *Instances of the Water,* borrowing the term from the hourglasses of the ancients, which contained water instead of sand. These measure nature by periods of time, as the instances of the rod by degrees of space. For all motion or natural action is performed in time, some more quickly, some more slowly, but all in periods determined and fixed in the nature of things. Even those actions which seem to be performed suddenly and (as we say) in the twinkling of an eye, are found to admit of degree in respect to duration.

First, then, we see that the revolutions of heavenly bodies are accomplished in calculated times, as also the flux and reflux of the sea. The motion of heavy bodies to the earth, and of light bodies toward the heavens, is accomplished in definite periods, varying with the bodies moved and the medium through which they move. The sailing of ships, the movements of animals, the transmission of missiles, are all performed likewise in times which admit (in the aggregate) of measurement. As for heat, we see boys in wintertime bathe their hands in flame without being burned, and jugglers by nimble and equable movements turn vessels full of wine or water upside down and then up again without spilling the liquid; and many other things of a similar kind. The compressions also and expansions and eruptions of bodies are performed, some more quickly, some more slowly, according to the nature of the body and motion, but in certain periods.

Moreover, in the explosion of several guns at once, which are heard sometimes to the distance of thirty miles, the sound is caught by those who are near the spot where the discharge is made sooner than by those who are at a greater distance. Even in sight, whereof the action is most rapid, it appears that there are required certain moments of time for its accomplishment, as is shown by those things which by reason of the velocity of their motion cannot be seen—as when a ball is discharged from a musket. For the ball flies past in less time than the image conveyed to the sight requires to produce an impression.

This fact, with others like it, has at times suggested to me a strange doubt, viz., whether the face of a clear and starlit sky be seen at the instant at which it really exists, and not a little later; and whether there be not, as regards our sight of heavenly bodies, a real time and an apparent time, just like the real place and apparent place which is taken account of by astronomers in the correction for parallaxes. So incredible did it appear to me that the images or rays of heavenly bodies could be conveyed at once to the sight through such an immense space and did not rather take a perceptible time in traveling to us. But this suspicion as to any considerable interval between the real time and the apparent afterward vanished entirely when I came to think of the infinite loss and diminution of quantity which distance causes in appearance between the real body of the star and its seen image; and at the same time when I observed the great distance (sixty miles at the least) at which bodies merely white are instantly seen here on earth; while there is no doubt that the light of heavenly bodies exceeds many times over in force of radiation not merely the vivid color of whiteness, but also the light of every flame that is known to us. Again, the immense velocity in the body itself as discerned in its daily motion (which has so astonished certain grave men that they preferred believing that the earth moved) renders this motion of ejaculation of rays therefrom (although wonderful, as I have said, in speed) more easy of belief. But what had most weight of all with me was that if any perceptible interval of time were interposed between the reality and the sight, it would follow that the images would oftentimes be intercepted and confused by clouds rising in the meanwhile, and similar disturbances in the medium. And thus much for the simple measures of time.

But not only must we seek the measure of motions and actions by themselves but much more in comparison, for this is of excellent use and very general application. Now we find that the flash of a gun is seen sooner than its report is heard, although the ball must necessarily strike the air before the flame behind it can get out. And this is owing, it seems, to the motion of light being more rapid than that of sound. We find, too, that visible images are received by the sight faster than they are dismissed. Thus the strings of a violin when struck by the finger are to appearance doubled or tripled, because a new image is received before the old one is gone; which is also the reason why rings being spun round look like globes, and a lighted torch, carried hastily at night, seems to have a tail. And it was upon this inequality of motions in point of velocity that Galileo built his theory of the flux and reflux of the sea, supposing that the earth revolved faster than the water could follow, and that the water therefore first gathered in a heap and then fell down, as we see it do in a basin of water moved quickly. But this he devised upon an assumption which cannot be allowed, viz., that the earth moves, and also without being well informed as to the sex-horary motion of the tide.

But an example of the thing I am treating of, to wit, the comparative measures of motions—and not only of the thing itself, but also of its eminent use (of which I spoke just now) —is conspicuous in mining with gunpowder where vast masses of earth, buildings, and the like are upset and thrown into the air by a very small quantity of powder. The cause of which is doubtless this: that the motion of expansion in the impelling powder is quicker many times over than the motion of the resisting gravity, so that the first motion is over before the countermotion is begun, and thus at first the resistance amounts to nothing. Hence too it happens that in projectiles it is not the strong blow but the sharp and quick that carries the body furthest. Nor would it be possible for the small quantity of animal spirit in animals, especially in such huge creatures as the whale or elephant, to bend and guide such a vast mass of body were it not for the velocity of the spirit's motion, and the slowness of the bodily mass in exerting its resistance.

This one thing indeed is a principal foundation of the experiments in natural magic (of which I shall speak presently) wherein a small mass of matter overcomes and regulates a far larger mass—I mean the contriving that of two motions one shall by its superior velocity get the start and take effect before the other has time to act.

Lastly, this distinction of foremost and hindmost ought to be observed in every natural action. Thus in an infusion of rhubarb the purgative virtue is extracted first, the astringent afterward. And something of the kind I have found on steeping violets in vinegar, where the sweet and delicate scent of the flower is extracted first, and then the more earthy part of the flower, which mars the scent. Therefore, if violets be steeped in vinegar for a whole day the scent is extracted much more feebly, but if you keep them in for a quarter of an hour only and then take them out, and (since the scented spirit in violets is small) put in fresh violets every quarter of an hour as many as six times, the infusion is at last so enriched that although there have not been violets in the vinegar, however renewed, for more than an hour and a half altogether, there nevertheless remains in it a most grateful odor, as strong as the violet itself, for an entire year. It should be observed, however, that the odor does not gather its full strength till after a month from the time of infusion. In the distillation too of aromatic herbs crushed in spirit of wine, it appears that there first rises an aqueous and useless phlegm, then a water containing more of the spirit of wine, and lastly, a water containing more of the aroma. And of this kind there are to be found in distillations a great many facts worthy of notice. But let these suffice for examples.

XLVII

Among Prerogative Instances I will put in the twenty-third place *Instances of Quantity,* which (borrowing a term from medicine) I also call *Doses of Nature.* These are they which measure virtues according to the *quantity* of the bodies in which they subsist and show how far the *mode* of the virtue depends upon the *quantity* of the body. And first there are certain virtues which subsist only in a cosmical quantity, that is, such a quantity as has consent with the configuration and fabric of the universe. The earth for instance stands fast; its parts fall. The waters in seas ebb and flow; but not in rivers, except through the sea coming up. Secondly, almost all particular virtues act according to the greater or less quantity of the body. Large quantities of water corrupt slowly, small ones quickly. Wine and beer ripen and become fit to drink much more quickly in bottles than in casks. If an herb be steeped in a large quantity of liquid, infusion takes place rather than impregnation; if in a small, impregnation rather than infusion. Thus in its effect on the human body a bath is one thing, a slight sprinkling another. Light dews, again, never fall in the air but are dispersed and incorporated with it. And in breathing on precious stones you may see the slight moisture instantly dissolved, like a cloud scattered by the wind. Once more, a piece of a magnet does not draw so much iron as the whole magnet. On the other hand there are virtues in which smallness of quantity has more effect, as in piercing, a sharp point pierces more quickly than a blunt one; a pointed diamond cuts glass, and the like.

But we must not stay here among indefinites, but proceed to inquire what *proportion* the quantity of a body bears to the mode of its virtue. For it would be natural to believe that the one was equal to the other; so that if a bullet of an ounce weight falls to the ground in a given time, a bullet of two ounces ought to fall twice as quickly, which is not the fact. Nor do the same proportions hold in all kinds of virtues, but widely different. These measures, therefore, must be sought from experiment, and not from likelihood or conjecture.

Lastly, in all investigation of nature the quantity of body—the dose, as it were—required to produce any effect must be set down, and cautions as to the too little and too much be interspersed.

XLVIII

Among Prerogative Instances I will put in the twenty-fourth place *Instances of Strife,* which I also call *Instances of Predominance.* These indicate the mutual predominance and subjection of virtues: which of them is stronger and prevails, which of them is weaker and gives way. For the motions and efforts of bodies are compounded, decomposed, and compli-cated, no less than the bodies themselves. I will therefore first propound the principal kinds of motions or active virtues in order that we may be able more clearly to compare them to-gether in point of strength, and thereby to point out and designate more clearly the instances of strife and predomi-nance.

Let the first motion be that motion of *resistance* in matter which is inherent in each several portion of it, and in virtue of which it absolutely refuses to be annihilated. So that no fire, no weight or pressure, no violence, no length of time can re-duce any portion of matter, be it ever so small, to nothing, but it will ever be something, and occupy some space; and, to whatever straits it may be brought, will free itself by chang-ing either its form or its place; or if this may not be, will sub-sist as it is; and will never come to such a pass as to be either nothing or nowhere. This motion the Schoolmen (who almost always name and define things rather by effects and incapaci-ties than by inner causes) either denote by the axiom "two bodies cannot be in one place," or call "the motion to prevent penetration of dimensions." Of this motion it is unnecessary to give examples, as it is inherent in every body.

Let the second motion be what I call motion of *connection*, by which bodies do not suffer themselves to be separated at any point from contact with another body, as delighting in mutual connection and contact. This motion the Schoolmen call "motion to prevent a vacuum," as when water is drawn up by suction or in a pump; the flesh by cupping glasses; or when water stops without running out in perforated jars unless the mouth of the jar be opened to let in the air; and in numberless instances of a similar kind.

Let the third motion be what I call motion of *liberty*, by which bodies strive to escape from preternatural pressure or tension and to restore themselves to the dimensions suitable to their nature. Of this motion also we have innumerable examples, such as (to speak first of escape from pressure) the motion of water in swimming, of air in flying, of water in rowing, of air in the undulations of winds, of a spring in clocks—of which we have also a pretty instance in the motion of the air compressed in children's popguns, when they hollow out an alder twig or some such thing and stuff it up at both ends with a piece of pulpy root or the like, and then with a ramrod thrust one of the roots or whatever the stuffing be toward the other hole, from which the root at the further end is discharged with a report, and that before it is touched by the nearer root or the ramrod. As for bodies escaping from tension, this motion displays itself in air remaining in glass eggs after suction; in strings, in leather and in cloth, which recoil after tension, unless it has gained too great strength by continuance; and in similar phenomena. This motion the Schoolmen refer to under the name of "motion in accordance with the form of the element"; an injudicious name enough, since it is a motion which belongs not only to fire, air, and water, but to every variety of solid substance, as wood, iron, lead, cloth, parchment, etc.; each of which bodies has its own proper limit of dimension out of which it cannot easily be drawn to any considerable extent. But since this motion of

liberty is of all the most obvious, and is of infinite application, it would be a wise thing to distinguish it well and
clearly. For some very carelessly confuse this motion with the
two former motions of resistance and connection, the motion,
that is, of escape from pressure with the motion of resistance;
of escape from tension with the motion of connection—just as
if bodies when compressed yield or expand, that there may
not ensue penetration of dimensions; and, when stretched, recoil and contract, that there may not ensue a vacuum. Whereas
if air when compressed had a mind to contract itself to the
density of water, or wood to the density of stone, there would
be no necessity for penetration of dimensions, yet there might
be a far greater compression of these bodies than they ever do
actually sustain. In the same way, if water had a mind to expand to the rarity of air, or stone to the rarity of wood, there
would be no need for a vacuum to ensue, and yet there
might be effected a far greater extension of these bodies than
they ever do actually sustain. Thus the matter is never
brought to a penetration of dimensions or to a vacuum, except in the extreme limits of condensation and rarefaction,
whereas the motions of which I speak stop far short of these
limits, and are nothing more than desires which bodies have
for preserving themselves in their consistencies (or, if the
Schoolmen like, in their forms), and not suddenly departing
therefrom unless they be altered by gentle means, and with
consent. But it is far more necessary (because much depends
upon it) that men should know that violent motion (which
we call mechanical, but which Democritus, who in expound-
ing his primary motions is to be ranked even below second-
rate philosophers, called motion of stripe) is nothing more
than this motion of liberty, that is, of escape from compression to relaxation. For either in a mere thrust, or in flight
through the air, there occurs no movement or change of place
until the parts of the body moved are acted upon and compressed by the impelling body more than their nature will
bear. Then, indeed, when each part pushes against the next,
one after the other, the whole is moved. And it not only

moves forward, but revolves at the same time, the parts seeking in that way also to free themselves or to distribute the pressure more equally. And so much for this motion.

Let the fourth motion be that to which I have given the name of the motion of *matter*, which is in some sort the converse of the last named motion. For in the motion of liberty bodies dread, loathe, and shun a new dimension, or a new sphere, or new expansion or contraction (which are all names for the same thing), and strive with all their might to recoil, and recover their old consistency. On the contrary, in this motion of matter bodies desire a new sphere or dimension and aspire thereto readily and quickly, and sometimes, as in the case of gunpowder, with most violent effort. Now the instruments of this motion, not indeed the sole, but the most potent, or at any rate the most common, are heat and cold. For instance, air, if expanded by tension, as by suction in glass eggs, labors under a strong desire to recover itself. But if heat be applied, it longs, on the contrary, to expand, and desires a new sphere and passes into it readily as into a new form (so they phrase it); and after a certain degree of expansion cares not to return, unless invited thereto by the application of cold, which is not a return, but a renewed transmutation. In the same way water, if made to contract by pressure, resists and wishes to become such as it was, that is, larger. But if there intervene intense and continued cold, it changes itself spontaneously and gladly to the density of ice; and if the cold be continued long, without interruption from heat, as in grottoes and caverns of some depth, it turns to crystal or some similar material and never recovers its form.

Let the fifth motion be the motion of *continuity*, by which I do not mean simple and primary continuity with some other body (for that is the motion of connection), but self-continuity in a given body. For it is most certain that all bodies dread a solution of continuity, some more, some less, but all to a certain extent. For while in hard bodies, as steel or glass, the resistance to discontinuity is exceedingly strong, even in liquids, where it seems to disappear or at all events to be very feeble, it is not altogether absent but is certainly there, though in its lowest degree of power, and betrays itself in very many experiments as in bubbles, in the roundness of drops, in the thin threads of droppings from roofs, in the tenacity of glutinous bodies, and the like. But most of all does this appetite display itself if an attempt be made to extend the discontinuity to minute fragments. For in a mortar, after a certain amount of pulverization, the pestle produces no further effect; water does not penetrate into minute chinks; even air itself, notwithstanding its subtlety, does not suddenly pass through the pores of solid vessels but only after long insinuation.

Let the sixth motion be that which I call motion *for gain*, or motion *of want*. It is that by which bodies, when placed among quite heterogeneous and hostile bodies, if they find an opportunity of escaping from these and uniting themselves to others more cognate (though these others be such as have no close union with them) do nevertheless embrace the latter and choose them as preferable; and seem to view this connection in the light of a *gain* (whence the term), as though they stood in need of such bodies. For instance, gold or any other metal in the leaf does not like the surrounding air. If therefore it meet with any thick tangible body (as a finger, paper, what you will) it instantly sticks to it and is not easily torn away. So too paper, cloth, and the like do not agree well with the air which is lodged in their pores. They are therefore glad to imbibe water or other moisture and eject the air. A piece of sugar too, or a sponge, if dipped at one end in water or wine, while the other stands out far above the surface, draws the water or the wine gradually upward.

Hence we derive an excellent rule for opening and dissolving bodies. For (to say nothing of corrosives and strong waters which open for themselves a way) if there can be found a body proportioned to and more in harmony and affinity with a given solid body than that with which it is as of necessity mixed, the solid body immediately opens and relaxes itself, and shutting out or ejecting the latter, receives the former into itself. Nor does this motion for gain act or exist only in immediate contact. For electricity (of which Gilbert and others after him have devised such stories) is nothing else than the appetite of a body when excited by gentle friction— an appetite which does not well endure the air but prefers some other tangible body, if it be found near at hand.

Let the seventh motion be what I call the motion *of the greater congregation*, by which bodies are carried toward masses of a like nature with themselves—heavy bodies to the globe of the earth, light to the compass of the heaven. This the Schoolmen have denoted by the name of *natural motion* from superficial considerations; either because there was nothing conspicuous externally which could produce such motion (and therefore they supposed it to be innate and inherent in things themselves), or perhaps because it never ceases. And no wonder; for the earth and heaven are ever there, whereas the causes and origins of most other motions are sometimes absent, sometimes present. Accordingly this motion, because it ceases not but when others cease is felt instantly, they deem perpetual and proper, all others adscititious. This motion, however, in point of fact is sufficiently weak and dull, being one which, except in bodies of considerable bulk, yields and succumbs to all other motions, as long as they are in operation. And though this motion has so filled men's thoughts as to have put all others almost out of sight, yet it is but little that they know about it, being involved in many errors with regard to it.

Let the eighth motion be the motion *of the lesser congregation*, by which the homogeneous parts in a body separate themselves from the heterogeneous and combine together; by which also entire bodies from similarity of substance embrace and cherish each other, and sometimes are attracted and collected together from a considerable distance; as when in milk, after it has stood a while, the cream rises to the top, while in wine the dregs sink to the bottom. For this is not caused by the motion of heaviness and lightness only, whereby some parts rise up and some sink down, but much more by a desire of the homogeneous parts to come together and unite in one.

Now this motion differs from the motion of want in two points. One is that in the latter there is the stronger stimulus of a malignant and contrary nature, whereas in this motion (provided there be nothing to hinder or fetter it) the parts unite from friendship even in the absence of a foreign nature to stir up strife. The other point is that the union is here closer and, as it were, with greater choice. In the former, if only the hostile body be avoided, bodies not closely related come together, whereas in the latter, substances are drawn together by the tie of close relationship and, as it were, combine into one. And this motion resides in all composite bodies and would readily show itself were it not bound and restrained by other appetites and necessities in the bodies which interfere with the union in question.

Now the binding of this motion takes place generally in three ways: by the torpor of bodies; by the check of a dominant body; and by external motions. Now, for the torpor of bodies, it is certain that there resides in tangible substances a certain sluggishness, more or less, and an aversion from change of place; insomuch that, unless they be excited, they had rather remain as they are than change for the better. Now this torpor is shaken off by the help of three things: either by heat, or by the eminent virtue of some cognate body, or by lively and powerful motion. And as for the help of heat, it is for this reason that heat has been defined to be "that which separates Heterogeneous and congregates Homogeneous parts"; a definition of the Peripatetics justly derided by Gilbert, who says it is much the same as if a man were to be defined as that which sows wheat and plants vines—for that it is, a definition simply by effects, and those particular. But the definition has a worse fault, inasmuch as these effects, such as they are, arise not from a peculiar property of heat, but only indirectly (for cold does the same, as I shall afterwards show); being caused by the desire of homogeneous parts to unite, heat simply aiding to shake off the torpor which had previously bound the desire. As for the help derived from the virtue of a cognate body, it is well seen in an armed magnet which excites in iron the virtue of detaining iron by similarity of substance, the torpor of the iron being cast off by the virtue of the magnet. And as for help derived from motion, it is shown in wooden arrows, having their points also of wood, which penetrate more deeply into wood than if they were tipped with steel, owing to the similarity of substance, the torpor of the wood being shaken off by the rapid motion. Of these two experiments I have spoken also in the Aphorism on Clandestine Instances.

That binding of the motion of the lesser congregation which is caused by the restraint of a dominant body is seen in the resolution of blood and urine by cold. For as long as those bodies are filled with the active spirit which, as lord of the whole, orders and restrains the several parts of whatsoever sort, so long the homogeneous parts do not meet together on account of the restraint. But as soon as the spirit has evaporated, or been choked by cold, then the parts being freed from restraint meet together in accordance with their natural desire. And thus it happens that all bodies which contain an eager spirit (as salts and the like) remain as they are, and are not resolved, owing to the permanent and durable restraint of a dominant and commanding spirit.

That binding of the motion of lesser congregation which is caused by external motion is most conspicuous in the shaking of bodies to prevent putrefaction. For all putrefaction depends on the assembling together of homogeneous parts, whence there gradually ensues the corruption of the old form, as they call it, and the generation of a new. For putrefaction, which paves the way for the generation of a new form, is preceded by a dissolution of the old, which is itself a meeting together of homogeneous parts. That, indeed, if not impeded, is simple resolution. But if it be met by various obstacles there follow putrefactions, which are the rudiments of a new generation. But if (which is the present question) a frequent agitation be kept up by external motion, then indeed this motion of uniting (which is a delicate and tender one, and requires rest from things without) is disturbed and ceases, as we see happen in numberless instances. For example, the daily stirring or flowing of water prevents it from putrefying; winds keep off pestilence in the air; corn turned and shaken in the granary remains pure; all things, in short, that are shaken outwardly are the slower to putrefy inwardly.

Lastly, I must not omit that meeting of the parts of bodies which is the chief cause of induration and desiccation. For when the spirit, or moisture turned to spirit, has escaped from some porous body (as wood, bone, parchment, and the like), then the grosser parts are with stronger effort drawn and collected together; whence ensues induration or desiccation, which I take to be owing not so much to the motion of connection to prevent a vacuum as to this motion of friendship and union.

As for the meeting of bodies from a distance, that is a rare occurrence, and yet it exists in more cases than are generally observed. We have illustrations of it when bubble dissolves bubble; when medicines draw humors by similarity of substance; when the chord of one violin makes the chord of another sound a unison, and the like. I suspect also that this motion prevails in the spirits of animals, though it be altogether unknown. At any rate it exists conspicuously in the magnet and magnetized iron. And now that we are speaking of the motions of the magnet, they ought to be carefully distinguished. For there are four virtues or operations in the magnet which should not be confounded but kept apart, although the wonder and admiration of men have mixed them up together. The first is, the attraction of magnet to magnet, or of iron to magnet, or of magnetized iron to iron. The second is its polarity, and at the same time its declination. The third, its power of penetrating through gold, glass, stone, everything. The fourth, its power of communicating its virtue from stone to iron, and from iron to iron, without communication of substance. In this place, however, I am speaking only of the first of these virtues—that is, its attractive power. Remarkable also is the motion of attraction between quicksilver and gold, insomuch that gold attracts quicksilver, though made up into ointments; and men who work amid the vapors of quicksilver usually hold a piece of gold in their mouths to collect the exhalations which would otherwise penetrate into their skulls and bones; by which also the piece of gold is presently turned white. And so much for the motion of the lesser congregation.

Let the ninth motion be the *magnetic,* which, though it be of the same genus with the motion of the lesser congregation, yet if it operates at great distances and on large masses, deserves a separate investigation, especially if it begin not with contact, as most, nor lead to contact, as all motions of congregation do, but simply raises bodies or makes them swell, and nothing more. For if the moon raises the waters, or makes moist things swell; if the starry heaven attracts planets to their apogees; if the sun holds Venus and Mercury so that their elongations never exceed a certain distance; these motions seem to fall properly neither under the greater nor the lesser congregation, but to be of a sort of intermediate and imperfect congregation, and therefore ought to constitute a species by themselves.

Let the tenth motion be that of *flight* (a motion the exact opposite of that of the lesser congregation), by which bodies from antipathy flee from and put to flight hostile bodies, and separate themselves from them or refuse to mingle with them. For although in some cases this motion may seem to be an accident or a consequence of the motion of the lesser congregation, because the homogeneous parts cannot meet without dislodging and ejecting the heterogeneous, still it is a motion that should be classed by itself and formed into a distinct species, because in many cases the appetite of flight is seen to be more dominant than the appetite of union.

This motion is eminently conspicuous in the excretions of animals and not less in objects odious to some of the senses, especially the smell and the taste. For a fetid odor is so rejected by the sense of smell as to induce by consent in the mouth of the stomach a motion of expulsion; a rough and bitter taste is so rejected by the palate or throat as to induce by consent a shaking of the head and a shudder. But this motion has place in other things also. It is observed in certain forms of reaction; as in the middle region of the air, where the cold seems to be the effect of the rejection of the nature of cold from the confines of the heavenly bodies; as also the great heats and burnings which are found in subterranean places appear to be rejections of the nature of heat from the inner parts of the earth. For heat and cold, in small quantities, kill one another. But if they be in large masses, and as it were in regular armies, the result of the conflict is that they displace and eject each other in turn. It is also said that cinnamon and other perfumes retain their scent longer when placed near sinks and foul-smelling places because they refuse to come out and mingle with stenches. It is certain that quicksilver, which of itself would reunite into an entire mass, is kept from doing so by spittle, hog's lard, turpentine, and the like, owing to the ill consent which its parts have with such bodies, from which, when spread around them, they draw back, so that their desire to fly from these intervening bodies is more powerful than their desire of uniting with parts like themselves. And this is called the *mortification* of quicksilver. The fact also that oil does not mix with water is not simply owing to the difference of weight, but to the ill consent of these fluids, as may be seen from the fact that spirit of wine, though lighter than oil, yet mixes well enough with water. But most of all is the motion of flight conspicuous in niter and such like crude bodies, which abhor flame; as in gunpowder, quicksilver, and gold. But the flight of iron from one pole of the magnet is well observed by Gilbert to be not a flight strictly speaking, but a conformity and meeting in a more convenient situation.

Let the eleventh motion be that of *assimilation,* or of *self-multiplication,* or again of simple *generation.* By which I mean not the generation of integral bodies, as plants or animals, but of bodies of uniform texture. That is to say, by this motion such bodies convert others which are related, or at any rate well disposed to them, into their own substance and nature. Thus flame over vapors and oily substances multiplies itself and generates new flame; air over water and watery substances multiplies itself and generates new air; spirit, vegetable and animal, over the finer parts as well of watery as of oily substance in its food, multiplies itself and generates new spirit; the solid parts of plants and animals, as the leaf, flower, flesh, bone, and the like, severally assimilate new substance to follow and supply what is lost out of the juices of their food. For let no one adopt the wild fancy of Paracelsus who (blinded I suppose by his distillations) will have it that nutrition is caused only by separation, and that in bread and meat lie eye, nose, brain, liver; in the moisture of the ground, root, leaf, and flower. For as the artist out of the rude mass of stone or wood educes, by separation and rejection of what is superfluous, leaf, flower, eye, nose, hand, foot, and the like, so, he maintains, does Archæus, the internal artist, educe out of food by separation and rejection the several members and parts of our body. But to leave such trifles, it is most certain that the several parts, as well similar as organic, in vegetables and animals do first attract with some degree of selection the juices of their food, which are alike or nearly so for all, and then assimilate them and turn them into their own nature. Nor does this assimilation or simple generation take place only in animate bodies, but inanimate also participate therein, as has been stated of flame and air. Moreover, the nonvital spirit, which is contained in every tangible animated substance, is constantly at work to digest the coarser parts and turn them into spirit, to be afterwards discharged; whence ensues diminution of weight and desiccation, as I have stated elsewhere. Nor must we set apart from assimilation that accretion which is commonly distinguished from alimentation; as when clay between stones concretes and turns into a stony substance, or the scaly substance on the teeth turns into a substance as hard as the teeth themselves, and so on. For I am of opinion that there resides in all bodies a desire for assimilation as well as for uniting with homogeneous substances; but this virtue is bound, as is the other, though not by the same means. But these means, as well as the way of escape from them, ought to be investigated with all diligence because they pertain to the rekindling of the vital power in old age. Lastly, it seems worthy of observation that in the nine motions of which I have spoken [1] bodies seem to desire only the preservation of their nature, but in this tenth the propagation of it.

Let the twelfth motion be that of *excitation*, a motion which seems to belong to the genus of assimilation and which I sometimes call by that name. For it is a motion diffusive, communicative, transitive, and multiplicative, as is the other, and agreeing with it generally in effect though differing in the mode of effecting and in the subject matter. For the motion of assimilation proceeds, as it were, with authority and command; it orders and forces the assimilated body to turn into the assimilating. But the motion of excitation proceeds, so to speak, with art and by insinuation, and stealthily, simply inviting and disposing the excited body to the nature of the exciting. Again, the motion of assimilation multiplies and transforms bodies and substances. Thus more flame is produced, more air, more spirit, more flesh. But in the motion of excitation virtues only are multiplied and transferred; more heat being engendered, more magnetic power, more putrefying. This motion is particularly conspicuous in heat and cold. For heat does not diffuse itself, in heating a body, by communication of the original heat but simply by exciting the parts of the body to that motion which is the form of heat, of which I have spoken in the First Vintage concerning the nature of heat. Consequently heat is excited far more slowly and with far greater difficulty in stone or metal than in air, owing to the unfitness and unreadiness of those bodies to receive the motion. So that it is probable that there may exist materials in the bowels of the earth which altogether refuse to be heated, because through their greater condensation they are destitute of that spirit with which this motion of excitation generally begins. In like manner the magnet endues iron with a new disposition of its parts and a conformable motion, but loses nothing of its own virtue. Similarly leaven,

[1] [which relate to concrete bodies rather than to matter in general —Ed.]

yeast, curd, and certain poisons excite and invite a successive and continued motion in dough, beer, cheese, or the human body, not so much by the force of the exciting as by the predisposition and easy yielding of the excited body.

Let the thirteenth motion be the motion of *impression,* which also is of the same genus with the motion of assimilation, and is of diffusive motions the most subtle. I have thought fit, however, to make a distinct species of it, on account of a remarkable difference between it and the two former. For the simple motion of assimilation actually transforms the bodies themselves, so that you may take away the first mover, and there will be no difference in what follows. For the first kindling into flame, or the first turning into air, has no effect on the flame or air next generated. In like manner, the motion of excitation continues, after the first mover is withdrawn, for a very considerable time: as in a heated body when the primary heat has been removed; in magnetized iron when the magnet has been put away; in dough when the leaven has been taken out. But the motion of impression, though diffusive and transitive, seems to depend forever on the prime mover. So that if that be taken away or cease to act, it immediately fails and comes to an end, and therefore the effect must be produced in a moment, or at any rate in a very brief space of time. The motions therefore of assimilation and excitation I call motions of the *generation of Jupiter,* because the generation continues; but this, the motion of the *generation of Saturn,* because the birth is immediately devoured and absorbed. It manifests itself in three things: in rays of light, in the percussions of sounds, and in magnetism, as regards the communication of the influence. For if you take away light, colors and its other images instantly disappear; if you take away the original percussion and the vibration of the body thence produced, the sound soon after dies away. For though sounds are troubled as they pass through their medium by winds, as if by waves, yet it must be carefully noted that the original sound does not last all the time the resonance goes on. For if you strike a bell, the sound seems to be continued for a good long time, whereby we might easily be led into the error of supposing that during the whole of the time the sound is, as it were, floating and hanging in the air, which is quite untrue. For the resonance is not the identical sound, but a renewal of it, as is shown by quieting or stopping the body struck. For if the bell be held tight so that it cannot move, the sound at once comes to an end and resounds no more—as in stringed instruments, if after the first percussion the string be touched, either with the finger, as in the harp, or with the quill, as in the spinet, the resonance immediately ceases. Again, when the magnet is removed, the iron immediately drops. The moon indeed cannot be removed from the sea, nor the earth from the falling body, and therefore we can try no experiment in these cases; but the principle is the same.

Let the fourteenth motion be the motion of *configuration* or *position*, by which bodies seem to desire not union or separation, but position, collocation, and configuration with respect to others. This motion is a very abstruse one and has not been well investigated. In some cases, indeed, it seems to be without a cause, though not, I believe, really so. For if it be asked why the heavens revolve rather from east to west than from west to east, or why they turn on poles placed near the Bears rather than about Orion, or in any other part of heaven, such questions seem to border on insanity, since these phenomena ought rather to be received as results of observation, and merely positive facts. But though there are no doubt in nature certain things ultimate and without cause, this does not appear to me to be one of them, being caused in my opinion by a certain harmony and consent of the universe which has not yet fallen under observation. And if we admit the motion of the earth from west to east, the same questions remain. For it also moves on certain poles. And why, it might be asked, should these poles be placed where they are, rather than anywhere else? Again the polarity, direction, and declination of the magnet are referable to this motion. There are also found in bodies natural as well as artificial, especially in solids, a certain collocation and position of parts, and a kind of threads and fibers, which ought to be carefully investigated since, until they are understood, these bodies cannot be conveniently managed or controlled. But those eddyings in fluids, by which when pressed, before they can free themselves, they relieve each other that they may all have a fair share of the pressure, belong more properly to the motion of liberty.

Let the fifteenth motion be the motion of *transition,* or motion *according to the passages,* by which the virtues of bodies are more or less impeded or promoted by their media, according to the nature of the body and of the acting virtues, and also of the medium. For one medium suits light, another sound, another heat and cold, another magnetic virtues, and so on.

Let the sixteenth motion be the *royal* (as I call it) or *political* motion, by which the predominant and commanding parts in any body curb, tame, subdue, and regulate the other parts, and compel them to unite, separate, stand still, move, and range themselves, not in accordance with their own desires, but as may conduce to the well-being of the commanding part; so that there is a sort of government and polity exerted by the ruling over the subject parts. This motion is eminently conspicuous in the spirits of animals where, as long as it is in vigor, it tempers all the motions of the other parts. It is found however in other bodies in a lower degree; as I said of blood and urine, which are not decomposed till the spirit which mixes and keeps together their parts be discharged or quenched. Nor is this motion confined to spirits, though in most bodies the spirits are masters owing to their rapid and penetrating motion. But in bodies of greater density and not filled with a lively and quickening spirit (such as there is in quicksilver and vitriol), the thicker parts are the masters, so that unless this yoke and restraint be by some expedient shaken off, there is very little hope of any new transformation of such bodies. But let no one suppose that I am forgetful of the point at issue, because while this series and distribution of motions tends to nothing else but the better investigation of their predominancy by instances of strife, I now make mention of predominancy among the motions themselves. For in describing this royal motion I am not treating of the predominancy of motions or virtues, but of the predominancy of parts in bodies; such being the predominancy which constitutes the peculiar species of motion in question.

Let the seventeenth motion be the *spontaneous motion of rotation,* by which bodies delighting in motion and favorably placed for it enjoy their own nature, and follow themselves, not another body, and court (so to speak) their own embraces. For bodies seem either to move without limit, or to remain altogether at rest, or to tend to a limit at which, according to their nature, they either revolve or rest. Those which are favorably placed, if they delight in motion, move in a circle, with a motion, that is, eternal and infinite. Those which are favorably placed, and abhor motion, remain at rest. Those which are not favorably placed move in a right line (as the shortest path) to consort with bodies of their own nature. But this motion of rotation admits of nine differences regarding 1. the center round which the bodies move; 2. the poles on which they move; 3. their circumference or orbit, according to their distance from the center; 4. their velocity, according to the greater or less rapidity of their rotation; 5. the course of their motion, as from east to west, or from west to east; 6. their declination from a perfect circle by spiral lines more or less distant from their center; 7. their declination from a perfect circle by spiral lines more or less distant from their poles; 8. the greater or lesser distance of these spirals from each other; 9. and lastly, the variation of the poles themselves, if they be movable; which, however, has nothing to do with rotation unless it be circular. This motion in common and long received opinion is looked upon as the proper motion of heavenly bodies, though there is a grave dispute with regard to it among some both of the ancients and of the moderns, who have attributed rotation to the earth. But a juster question perhaps arises upon this (if it be not past question), namely, whether this motion (admitting that the earth stands still) is confined to the heavens, and does not rather descend and communicate itself to the air and waters. The motion of rotation in missiles, as in darts, arrows, musket balls, and the like, I refer to the motion of liberty.

Let the eighteenth motion be the motion of *trepidation,* to which, as understood by astronomers, I do not attach much credit. But in searching carefully everywhere for the appetites of natural bodies this motion comes before us and ought, it seems, to constitute a species by itself. It is a motion of what may be called perpetual captivity and occurs when bodies that have not quite found their right place, and yet are not altogether uneasy, keep forever trembling and stirring themselves restlessly, neither content as they are nor daring to advance further. Such a motion is found in the heart and pulses of animals, and must of necessity occur in all bodies which so exist in a mean state between conveniences and inconveniences that when disturbed they strive to free themselves, and being again repulsed, are yet forever trying again.

Let the nineteenth and last motion be one which, though it hardly answers to the name, is yet indisputably a motion; and let us call it the motion of *repose,* or of *aversion to move.* It is by this motion that the earth stands still in its mass while its extremities are moving toward the middle—not to an imaginary center, but to union. By this appetite also all bodies of considerable density abhor motion. Indeed, the desire of not moving is the only appetite they have; and though in countless ways they be enticed and challenged to motion, they yet, as far as they can, maintain their proper nature. And if compelled to move, they nevertheless seem always intent on recovering their state of rest and moving no more. While thus engaged, indeed, they show themselves active and struggle for it with agility and swiftness enough, as weary and impatient of all delay. Of this appetite but a partial representation can be seen, since here with us, from the subduing and concocting power of the heavenly bodies, all tangible substances are not only not condensed to their utmost, but are even mixed with some portion of spirit.

Thus, then, have I set forth the species or simple elements of motions, appetites, and active virtues, which are in nature most general. And under these heads no small portion of natural science is sketched out. I do not, however, mean to say that other species may not be added, or that the divisions I have made may not be drawn more accurately according to the true veins of nature, or reduced to a smaller number. Observe, nevertheless, that I am not here speaking of any abstract divisions, as if one were to say that bodies desire either the exaltation or the propagation or the fruition of their nature; or again, that the motions of things tend to the preservation and good either of the universe, as resistance and connection; or of great wholes, as the motions of the greater congregation, rotation, and aversion to move; or of special forms, as the rest. For though these assertions be true, yet unless they be defined by true lines in matter and the fabric of nature, they are speculative and of little use.   Meanwhile, these will suffice and be of good service in weighing the predominancies of virtues and finding out instances of strife, which is our present object.

For of the motions I have set forth some are quite invincible; some are stronger than others, fettering, curbing, arranging them; some carry farther than others; some outstrip others in speed; some cherish, strengthen, enlarge, and accelerate others.

The motion of resistance is altogether adamantine and invincible. Whether the motion of connection be so, I am still undecided. For I am not prepared to say for certain whether or no there be a vacuum, either collected in one place or interspersed in the pores of bodies. But of one thing I am satisfied, that the reason for which a vacuum was introduced by Leucippus and Democritus (namely, that without it the same bodies could not embrace and fill sometimes larger and sometimes smaller spaces) is a false one. For matter is clearly capable of folding and unfolding itself in space, within certain limits, without the interposition of a vacuum; nor is there in air two thousand times as much of vacuity as there is in gold, which on their hypothesis there should be. Of this I am sufficiently convinced by the potency of the virtues of pneumatical bodies (which otherwise would be floating in empty space like fine dust) and by many other proofs. As for the other motions, they rule and are ruled in turn, in proportion to their vigor, quantity, velocity, force of projection, and also to the helps and hindrances they meet with.

For instance, there are some armed magnets that hold and suspend iron of sixty times their own weight, so far does the motion of the lesser prevail over the motion of the greater congregation; but if the weight be increased, it is overcome. A lever of given strength will raise a given weight, so far does the motion of liberty prevail over that of the greater congregation; but if the weight be increased, it is overcome. Leather stretches to a certain extent without breaking, so far does the motion of continuity prevail over the motion of tension; but if the tension be increased, the leather breaks and the motion of continuity is overcome. Water runs out at a crack of a certain size, so far does the motion of the greater congregation prevail over the motion of continuity; but if the crack be smaller, it gives way, and the motion of continuity prevails. If you charge a gun with ball and sulphur only, and apply the match, the ball is not discharged, the motion of the greater congregation overcoming in this case the motion of matter. But if you charge with gunpowder, the motion of matter in the sulphur prevails, being aided by the motions of matter and of flight in the niter. And so of other cases. Instances of strife, therefore, which point out the predominancies of virtues together with the manner and proportion in which they predominate or give place, should be sought and collected from all quarters with keen and careful diligence.

Nor should we examine less carefully the modes in which these motions give way. That is to say, whether they stop altogether or whether they continue to resist but are overpowered. For in bodies here with us there is no real rest, either in wholes or in parts, but only in appearance. And this apparent rest is caused either by equilibrium, or by absolute predominancy of motions: by equilibrium, as in scales, which stand still if the weights be equal; by predominancy, as in watering pots with holes in them, where the water rests and is kept from falling out by the predominancy of the motion of connection. But it should be observed, as I have said, how far these yielding motions carry their resistance. For if a man be pinned to the ground, tied hand and foot, or otherwise held fast, and yet struggle to rise with all his might, the resistance is not the less though it be unsuccessful. But the real state of the case (I mean whether by predominancy the yielding motion is, so to speak, annihilated, or rather whether a resistance is continued, though we cannot see it) will perhaps, though latent in the conflicts of motions, be apparent in their concurrence. For example, let trial be made in shooting. See how far a gun will carry a ball straight, or as they say point-blank, and then try whether, if it be fired upward, the stroke will be feebler than when it is fired downward, where the motion of gravity concurs with the blow.

Lastly, such canons of predominance as we meet with should be collected; for instance, that the more common the good sought, the stronger the motion. Thus the motion of connection, which regards communion with the universe, is stronger than the motion of gravity, which regards only communion with dense bodies. Again, that appetites which aim at a private good seldom prevail against appetites which aim at a more public good, except in small quantities—rules which I wish held good in politics.

XLIX

Among Prerogative Instances I will put in the twenty-fifth place *intimating instances,* those, I mean, which intimate or point out what is useful to man. For mere power and mere knowledge exalt human nature, but do not bless it. We must therefore gather from the whole store of things such as make most for the uses of life. But a more proper place for speaking of these will be when I come to treat of applications to practice. Besides, in the work itself of interpretation in each particular subject, I always assign a place to the *human chart,* or *chart of things to be wished for.* For to form judicious wishes is as much a part of knowledge as to ask judicious questions.

L

Among Prerogative Instances I will put in the twenty-sixth place *Polychrest Instances,* or *Instances of General Use.* They are those which relate to a variety of cases and occur frequently and therefore save no small amount of labor and fresh demonstration. Of the instruments and contrivances themselves the proper place for speaking will be when I come to speak of applications to practice and modes of experimenting. Moreover, those which have been already discovered and come into use will be described in the particular histories of the several arts. At present I will subjoin a few general remarks on them as examples merely of this general use.

Besides the simple bringing together and putting asunder of them, man operates upon natural bodies chiefly in seven ways, viz., either by exclusion of whatever impedes and disturbs; or by compressions, extensions, agitations, and the like; or by heat and cold; or by continuance in a suitable place; or by the checking and regulation of motion; or by special sympathies; or by the seasonable and proper alternation, series, and succession of all these ways, or at any rate of some of them.

With regard to the first, the common air, which is everywhere about us and pressing in, and the rays of the heavenly bodies, cause much disturbance. Whatever therefore serves to exclude them may justly be reckoned among things of general use. To this head belong the material and thickness of the vessels in which the bodies are placed on which we are going to operate; also the perfect stopping up of vessels by consolidation and *lutum sapientiæ,* as the chemists call it. Also the closing in of substances by liquids poured on the outside is a thing of very great use, as when they pour oil on wine or juices of herbs, which spreading over the surface like a lid preserves them excellently from the injury of the air. Nor are powders bad things; for though they contain air mixed up with them, they yet repel the force of the body of air round about, as we see in the preservation of grapes and other fruits in sand and flour. It is good too to spread bodies over with wax, honey, pitch, and like tenacious substances, for the more perfect enclosure of them and to keep off the air and heavenly bodies. I have sometimes tried the effect of laying up a vessel or some other body in quicksilver, which of all substances that can be poured round another is far the densest. Caverns, again, and subterraneous pits are of great use in keeping off the heat of the sun and that open air which preys upon bodies, and such are used in the north of Germany as granaries. The sinking of bodies in water has likewise the same effect, as I remember to have heard of bottles of wine being let down into a deep well to cool, but through accident or neglect being left there for many years, and then taken out; and that the wine not only was free from sourness or flatness, but tasted much finer, owing, it would seem, to a more exquisite commixture of its parts. And if the case require that bodies be let down to the bottom of the water, as in a river or the sea, without either touching the water or being enclosed in stopped vessels, but surrounded by air alone, there is good use in the vessel which has been sometimes employed for working under water on sunk ships whereby divers are enabled to remain a long while below, and take breath from time to time. This machine was a hollow bell made of metal which, being let down parallel to the surface of the water, carried with it to the bottom all the air it contained. It stood on three feet (like a tripod) the height of which was somewhat less than that of a man, so that the diver, when his breath failed, could put his head into the hollow of the bell, take breath, and then go on with his work. I have heard also of a sort of machine or boat capable of carrying men under water for some distance. Be that as it may, under such a vessel as I have described bodies of any sort can easily be suspended, and it is on that account that I have mentioned this experiment.

There is also another advantage in the careful and complete closing of bodies. For not only does it keep the outer air from getting in (of which I have already spoken), but also it keeps the spirit of the body, on which the operation is going on inside, from getting out. For it is necessary for one who operates on natural bodies to be certain of his total quantities, that is, that nothing evaporates or flows away. For then and then only are profound alterations made in bodies when, while nature prevents annihilation, art prevents also the loss or escape of any part. On this subject there has prevailed a false opinion which, if true, would make us well nigh despair of preserving the perfect quantity without diminution, namely, that the spirits of bodies, and air when rarefied by a high degree of heat, cannot be contained in closed vessels but escape through their more delicate pores. To this opinion men have been led by common experiment of an inverted cup placed on water with a candle in it or a piece of paper lighted; the consequence of which is that the water is drawn up; and also by the similar experiment of cupping glasses which when heated over flame draw up the flesh. For in each of these experiments they imagine that the rarefied air escapes, and that its quantity being thereby diminished, the water or flesh comes up into its place by the motion of connection. But this is altogether a mistake. For the air is not diminished in quantity, but contracted in space; nor does the motion of the rising of the water commence till the flame is extinguished or the air cooled. And therefore physicians, to make their cupping glasses draw better, lay on them cold sponges dipped in water. And therefore there is no reason why men should be much afraid of the easy escape of air or spirits. For though it be true that the most solid bodies have pores, still air or spirit do not easily submit to such extremely fine comminution, just as water refuses to run out at very small chinks.

With regard to the second of the seven modes of operating above mentioned, it is particularly to be observed that compression and such violent means have indeed, with respect to local motion and the like, a most powerful effect, as in machines and projectiles, an effect which even causes the destruction of organic bodies and of such virtues as consist altogether in motion. For all life, nay all flame and ignition, is destroyed by compression, just as every machine is spoiled or deranged by the same. It causes the destruction likewise of virtues which consist in the position and coarser dissimilarity of parts. This is the case with colors, for the whole flower has not the same color as when it is bruised, nor the whole piece of amber as the same piece pulverized. So also it is with tastes. For there is not the same taste in an unripe pear as there is in a squeezed and softened one, for it manifestly contracts sweetness by the process. But for the more remarkable transformations and alterations of bodies of uniform structure such violent means are of little avail, since bodies do not acquire thereby a new consistency that is constant and quiescent, but one that is transitory and ever striving to recover and liberate itself. It would not be amiss, however, to make some careful experiments for the purpose of ascertaining whether the condensation or the rarefaction of a body of nearly uniform structure (as air, water, oil, and the like), being induced by violence, can be made to be constant and fixed, and to become a kind of nature. This should first be tried by simple continuance, and then by means of helps and consents. And the trial might easily have been made (if it had but occurred to me) when I was condensing water, as mentioned above, by hammer and press, till it burst forth from its enclosure. For I should have left the flattened sphere to itself for a few days, and after that drawn off the water, that so I might have seen whether it would immediately occupy the same dimensions which it had before condensation. If it had not done so, either immediately or at any rate soon after, we might have pronounced the condensation a constant one; if it had, it would have appeared that a restoration had taken place and that the compression was transitory. Something of a similar kind I might have tried also with the expansion of air in the glass eggs. For after powerful suction I might have stopped them suddenly and tightly; I might have left the eggs so stopped for some days and then tried whether on opening the hole the air would be drawn up with a hissing noise, or whether on plunging them into water, as much water would be drawn up as there would have been at first without the delay. For it is probable—at least it is worth trying—that this might have been, and may be, the case; since in bodies of structure not quite so uniform the lapse of time does produce such effects. For a stick bent for some time by compression does not recoil, and this must not be imputed to any loss of quantity in the wood through the lapse of time, since the same will be the case with a plate of steel if the time be increased, and steel does not evaporate. But if the experiment succeed not with mere continuance, the business must not be abandoned, but other aids must be employed. For it is no small gain if by the application of violence we can communicate to bodies fixed and permanent natures. For thus air can be turned into water by condensation, and many other effects of the kind can be produced, man being more the master of violent motions than of the rest.

The third of the seven modes above-mentioned relates to that which, whether in nature or in art, is the great instrument of operation, viz., heat and cold. And herein man's power is clearly lame on one side. For we have the heat of fire which is infinitely more potent and intense than the heat of the sun as it reaches us, or the warmth of animals. But we have no cold save such as is to be got in wintertime, or in caverns, or by application of snow and ice, which is about as much perhaps in comparison as the heat of the sun at noon in the torrid zone, increased by the reflections of mountains and walls. For such heat as well as such cold can be endured by animals for a short time. But they are nothing to be compared to the heat of a burning furnace, or with any cold corresponding to it in intensity. Thus all things with us tend to rarefaction, and desiccation, and consumption; nothing hardly to condensation and inteneration except by mixtures and methods

that may be called spurious. Instances of cold therefore should
be collected with all diligence. And such it seems may be
found by exposing bodies on steeples in sharp frosts; by laying them in subterranean caverns; by surrounding them with
snow and ice in deep pits dug for the purpose; by letting them
down into wells; by burying them in quicksilver and metals;
by plunging them into waters which petrify wood; by burying
them in the earth, as the Chinese are said to do in the making of porcelain, where masses made for the purpose are left,
we are told, underground for forty or fifty years, and transmitted to heirs, as a kind of artificial minerals; and by similar
processes. And so too all natural condensations caused by cold
should be investigated, in order that, their causes being
known, they may be imitated by art. Such we see in the
sweating of marble and stones; in the dews condensed on the
inside of windowpanes toward morning after a night's frost;
in the formation and gathering of vapors into water under
the earth, from which springs often bubble up. Everything of
this kind should be collected.

Besides things which are cold to the touch, there are found
others having the power of cold, which also condense, but
which seem to act on the bodies of animals only, and hardly
on others. Of this sort we have many instances in medicines
and plasters, some of which condense the flesh and tangible
parts, as astringent and inspissatory medicaments; while others
condense the spirits, as is most observable in soporifics. There
are two ways in which spirits are condensed by medicaments
soporific, or provocative of sleep: one by quieting their motion, the other by putting them to flight. Thus violets, dried
rose leaves, lettuce, and like benedict or benignant medicaments, by their kindly and gently cooling fumes invite the
spirits to unite and quiet their eager and restless motion.
Rose water, too, applied to the nose in a fainting fit, causes
the resolved and too relaxed spirits to recover themselves and,
as it were, cherishes them. But opiates and kindred medicaments put the spirits utterly to flight by their malignant and
hostile nature. And therefore if they be applied to an external

part, the spirits immediately flee away from that part and do not readily flow into it again; if taken internally, their fumes, ascending to the head, disperse in all directions the spirits contained in the ventricles of the brain; and these spirits thus withdrawing themselves, and unable to escape into any other part, are by consequence brought together and condensed, and sometimes are utterly choked and extinguished; though on the other hand these same opiates taken in moderation do by a secondary accident (namely, the condensation which succeeds the coming together) comfort the spirits and render them more robust, and check their useless and inflammatory motions; whereby they contribute no little to the cure of diseases and prolongation of life.

Nor should we omit the means of preparing bodies to receive cold. Among others I may mention that water slightly warm is more easily frozen than quite cold.

Besides, since nature supplies cold as sparingly, we must do as the apothecaries do who, when they cannot get a simple, take its succedaneum or *quid pro quo,* as they call it—such as aloes for balsam, cassia for cinnamon. In like manner we should look round carefully to see if there be anything that will do instead of cold, that is to say, any means by which condensations can be effected in bodies otherwise than by cold, the proper office of which is to effect them. Such condensations, as far as yet appears, would seem to be limited to four. The first of these is caused by simple compression, which can do but little for permanent density, since bodies recoil, but which perhaps may be of use as an auxiliary. The second is caused by the contraction of the coarser parts in a body after the escape of the finer, such as takes place in indurations by fire, in the repeated quenchings of metals, and like processes. The third is caused by the coming together of those homogeneous parts in a body which are the most solid, and which previously had been dispersed and mixed with the less solid; as in the restoration of sublimated mercury, which occupies a far greater space in powder than as simple mercury, and similarly in all purging of metals from their dross. The fourth is brought about through sympathy, by applying substances which from some occult power condense. These sympathies or consents at present manifest themselves but rarely, which is no wonder, since before we succeed in discovering forms and configurations we cannot hope for much from an inquiry into sympathies. With regard to the bodies of animals, indeed, there is no doubt that there are many medicines, whether taken internally or externally, which condense as it were by consent, as I have stated a little above. But in the case of in-animate substances such operation is rare. There has indeed been spread abroad, as well in books as in common rumor, the story of a tree in one of the Tercera or Canary Isles (I do not well remember which) which is constantly dripping, so as to some extent to supply the inhabitants with water. And Paracelsus says that the herb called *Ros Solis* is at noon and under a burning sun filled with dew, while all the other herbs round it are dry. But both of these stories I look upon as fabulous. If they were true, such instances would be of most signal use and most worthy of examination. Nor do I con-ceive that those honeydews, like manna, which are found on the leaves of the oak in the month of May, are formed and condensed by any peculiar property in the leaf of the oak, but while they fall equally on all leaves, they are retained on those of the oak as being well united and not spongy as most of the others are.

As regards heat, man indeed has abundant store and com-mand thereof, but observation and investigation are wanting in some particulars, and those the most necessary, let the al-chemists say what they will. For the effects of intense heat are sought for and brought into view, but those of a gentler heat, which fall in most with the ways of nature, are not explored and therefore are unknown. And therefore we see that by the heats generally used the spirits of bodies are greatly exalted, as in strong waters and other chemical oils; that the tangible parts are hardened and, the volatile being discharged, some-times fixed; that the homogeneous parts are separated, while the heterogeneous are in a coarse way incorporated and mixed up together; above all, that the junctures of composite bodies and their more subtle configurations are broken up and confounded. Whereas the operations of a gentler heat ought to have been tried and explored, whereby more subtle mixtures and regular configurations might be generated and educed, after the model of nature and in imitation of the works of the sun—as I have shadowed forth in the Aphorism on Instances of Alliance. For the operations of nature are performed by far smaller portions at a time, and by arrangements far more exquisite and varied than the operations of fire, as we use it now. And it is then that we shall see a real increase in the power of man when by artificial heats and other agencies the works of nature can be represented in form, perfected in virtue, varied in quantity, and, I may add, accelerated in time. For the rust of iron is slow in forming, but the turning into *Crocus Martis* is immediate; and it is the same with verdigris and ceruse; crystal is produced by a long process, while glass is blown at once; stones take a long time to grow, while bricks are quickly baked. Meanwhile (to come to our present business), heats of every kind, with their effects, should be diligently collected from all quarters and investigated—the heat of heavenly bodies by their rays direct, reflected, refracted, and united in burning glasses and mirrors; the heat of lightning, of flame, of coal fire; of fire from different materials; of fire close and open, straitened and in full flow, modified in fine by the different structures of furnaces; of fire excited by blowing; of fire quiescent and not excited; of fire removed to a greater or less distance; of fire passing through various media; moist heats, as of a vessel floating in hot water, of dung, of external and internal animal warmth, of confined hay; dry heats, as of ashes, lime, warm sand; in short, heats of all kinds with their degrees.

But above all we must try to investigate and discover the effects and operations of heat when applied and withdrawn gradually, orderly, and periodically, at due distances and for due times. For such orderly inequality is in truth the daughter of the heavens and mother of generation; nor is anything great to be expected from a heat either vehement or precipitate or that comes by fits and starts. In vegetables this is most manifest; and also in the wombs of animals there is a great inequality of heat, from the motion, sleep, food, and passions of the female in gestation. Lastly, in the wombs of the earth itself, those I mean in which metals and fossils are formed, the same inequality has place and force. Which makes the unskillfulness of some alchemists of the reformed school all the more remarkable—who have conceived that by the equable warmth of lamps and the like, burning uniformly, they can attain their end. And so much for the operations and effects of heat. To examine them thoroughly would be premature, till the forms of things and the configurations of bodies have been further investigated and brought to light. For it will then be time to seek, apply, and adapt our instruments when we are clear as to the pattern.

The fourth mode of operating is by continuance, which is as it were the steward and almoner of nature. Continuance I call it when a body is left to itself for a considerable time, being meanwhile defended from all external force. For then only do the internal motions exhibit and perfect themselves when the extraneous and adventitious are stopped. Now the works of time are far subtler than those of fire. For wine cannot be so clarified by fire as it is by time; nor are the ashes produced by fire so fine as the dust into which substances are resolved and wasted by ages. So too the sudden incorporations and mixtures precipitated by fire are far inferior to those which are brought about by time. And the dissimilar and varied configurations which bodies by continuance put on, such as putrefactions, are destroyed by fire or any violent heat. Meanwhile it would not be out of place to observe that the motions of bodies when quite shut up have in them something of violence. For such imprisonment impedes the spontaneous motions of the body. And therefore continuance in an open vessel is best for separations; in a vessel quite closed for commixtures; in a vessel partly closed, but with the air entering, for putrefactions. But, indeed, instances showing the effects and operations of continuance should be carefully collected from all quarters.

The regulation of motion (which is the fifth mode of operating) is of no little service. I call it regulation of motion when one body meeting another impedes, repels, admits or directs its spontaneous motion. It consists for the most part in the shape and position of vessels. Thus the upright cone in alembics helps the condensation of vapors; the inverted cone in receivers helps the draining off of the dregs of sugar. Sometimes a winding form is required, and one that narrows and widens in turn, and the like. For all percolation depends on this, that the meeting body opens the way to one portion of the body met and shuts it to another. Nor is the business of percolation or other regulation of motion always performed from without. It may also be done by a body within a body, as when stones are dropped into water to collect the earthy parts; or when syrups are clarified with the whites of eggs that the coarser parts may adhere thereto, after which they may be removed. It is also to this regulation of motion that Telesius has rashly and ignorantly enough attributed the shapes of animals, which he says are owing to the channels and folds in the womb. But he should have been able to show the like formation in the shells of eggs, in which there are no wrinkles or inequalities. It is true, however, that the regulation of motion gives the shapes in molding and casting.

Operations by consents or aversions (which is the sixth mode) often lie deeply hid. For what are called occult and specific properties, or sympathies and antipathies, are in great part corruptions of philosophy. Nor can we have much hope of discovering the consents of things before the discovery of forms and simple configurations. For consent is nothing else than the adaptation of forms and configurations to each other.

The broader and more general consents of things are not, however, quite so obscure. I will therefore begin with them. Their first and chief diversity is this, that some bodies differ widely as to density and rarity but agree in configurations, while others agree as to density and rarity but differ in configurations. For it has not been ill observed by the chemists in their triad of first principles that sulphur and mercury run through the whole universe. (For what they add about salt is absurd, and introduced merely to take in bodies earthy, dry, and fixed.) But certainly in these two one of the most general consents in nature does seem to be observable. For there is consent between sulphur, oil, and greasy exhalation, flame, and perhaps the body of a star. So is there between mercury, water and watery vapors, air, and perhaps the pure and intersidereal ether. Yet these two quaternions or great tribes of things (each within its own limits) differ immensely in quantity of matter and density, but agree very well in configuration; as appears in numerous cases. On the other hand metals agree well together in quantity and density, especially as compared with vegetables, etc., but differ very widely in configuration; while in like manner vegetables and animals vary almost infinitely in their configurations, but in quantity of matter or density their variation is confined to narrow limits.

The next most general consent is that between primary bodies and their supports, that is, their menstrua and foods. We must therefore inquire, under what climates, in what earth, and at what depth, the several metals are generated; and so of gems, whether produced on rocks or in mines; also in what soil the several trees and shrubs and herbs thrive best and take, so to speak, most delight; moreover what manurings, whether by dung of any sort, or by chalk, sea sand, ashes, etc., do the most good; and which of them are most suitable and effective according to the varieties of soil. Again, the grafting and inoculating of trees and plants, and the principle of it, that is to say, what plants prosper best on what stocks, depends much on sympathy. Under this head it would be an agreeable experiment, which I have heard has been lately tried, of engrafting forest trees (a practice hitherto confined to fruit trees), whereby the leaves and fruit are greatly enlarged and the trees made more shady. In like manner the different foods of animals should be noted under general heads, and with their negatives. For carnivorous animals cannot live on herbs, whence the order of Feuillans (though the will in man has more power over the body than in other animals) has after trial (they say) well nigh disappeared, the thing not being endurable by human nature. Also the different materials of putrefaction, whence animalculae are generated, should be observed.

The consents of primary bodies with their subordinates (for such those may be considered which I have noted) are sufficiently obvious. To these may be added the consents of the senses with their objects. For these consents, since they are most manifest and have been well observed and keenly sifted, may possibly shed great light on other consents also which are latent.

But the inner consents and aversions, or friendships and enmities, of bodies (for I am almost weary of the words sympathy and antipathy on account of the superstitions and vanities associated with them) are either falsely ascribed, or mixed with fables, or from want of observation very rarely met with. For if it be said that there is enmity between the vine and colewort, because when planted near each other they do not thrive, the reason is obvious—that both of these plants are succulent and exhaust the ground, and thus one robs the other. If it be said that there is consent and friendship between corn and the corn cockle or the wild poppy, because these herbs hardly come up except in ploughed fields, it should rather be said that there is enmity between them, because the poppy and corn cockle are emitted and generated from a juice of the earth which the corn has left and rejected; so that sowing the ground with corn prepares it for their growth. And of such false ascriptions there is a great number. As for fables, they should be utterly exterminated. There remains indeed a scanty store of consents which have been approved by sure experiment, such as those of the magnet and iron, of gold and quicksilver, and the like. And in chemical experiments on metals there are found also some others worthy of observation. But they are found in greatest abundance (if one may speak of abundance in such a scarcity) in certain medicines which by their occult (as they are called) and specific properties have relation either to limbs, or humors, or diseases, or sometimes to individual natures. Nor should we omit the consents between the motions and changes of the moon and the affections of bodies below, such as may be gathered and admitted, after strict and honest scrutiny, from experiments in agriculture, navigation, medicine, and other sciences. But the rarer all the instances of more secret consents are, the greater the diligence with which they should be sought after, by means of faithful and honest traditions and narrations; provided this be done without any levity or credulity, but with an anxious and (so to speak) a doubting faith. There remains a consent of bodies, inartificial perhaps in mode of operation, but in use a polychrest, which should in no wise be omitted, but examined into with careful attention. I mean the proneness or reluctance of bodies to draw together or unite by composition or simple apposition. For some bodies are mixed together and incorporated easily, but others with difficulty and reluctance. Thus powders mix best with water, ashes and lime with oils, and so on. Nor should we merely collect instances of the propensity or aversion of bodies for mixture, but also of the collocation of their parts, of their distribution and digestion when they are mixed, and finally of their predominancy after the mixture is completed.

There remains the seventh and last of the seven modes of operation, namely, the means of operating by the alternation of the former six. But it would not be seasonable to bring forward examples of this till our search has been carried somewhat more deeply into the others singly. Now a series or chain of such alternations, adapted to particular effects, is a thing at once most difficult to discover and most effective to work with. But men are utterly impatient both of the inquiry and the practice, though it is the very thread of the labyrinth as regards works of any magnitude. Let this suffice to exemplify the polychrest instances.

LI

Among Prerogative Instances I will put in the twenty-seventh and last place *Instances of Magic,* by which I mean those wherein the material or efficient cause is scanty or small as compared with the work and effect produced. So that even where they are common they seem like miracles; some at first sight, others even after attentive consideration. These, indeed, nature of herself supplies sparingly, but what she may do when her folds have been shaken out, and after the discovery of forms and processes and configurations, time will show. But these magical effects (according to my present conjecture) are brought about in three ways: either by self-multiplication, as in fire, and in poisons called specific, and also in motions which are increased in power by passing from wheel to wheel; or by excitation or invitation in another body, as in the magnet, which excites numberless needles without losing any of its virtue, or in yeast and the like; or by anticipation of motion, as in the case already mentioned of gunpowder and cannons and mines. Of which ways the two former require a knowledge of consents, the third a knowledge of the measurement of motions. Whether there be any mode of changing bodies *per minima* (as they call it) and of transposing the subtler configurations of matter (a thing required in every sort of transformation of bodies) so that art may be enabled to do in a short time that which nature accomplishes by many windings, is a point on which I have at present no sure indications. And as in matters solid and true I aspire to the ultimate and supreme, so do I forever hate all things vain and tumid, and do my best to discard them.

LII

So much then for the dignities or prerogatives of instances. It must be remembered, however, that in this Organon of mine I am handling logic, not philosophy. But since my logic

aims to teach and instruct the understanding, not that it may
with the slender tendrils of the mind snatch at and lay hold of
abstract notions (as the common logic does), but that it may
in very truth dissect nature, and discover the virtues and actions of bodies, with their laws as determined in matter; so
that this science flows not merely from the nature of the
mind, but also from the nature of things—no wonder that it
is everywhere sprinkled and illustrated with speculations and
experiments in nature, as examples of the art I teach. It appears then from what has been said that there are twenty-
seven prerogative instances, namely, solitary instances; migratory instances; striking instances; clandestine instances; constitutive instances; conformable instances; singular instances;
deviating instances; bordering instances; instances of power;
instances of companionship and of enmity; subjunctive in-
stances; instances of alliance; instances of the fingerpost; in-
stances of divorce; instances of the door; summoning instances;
instances of the road; instances supplementary; dissecting instances; instances of the rod; instances of the course; doses of
nature; instances of strife; intimating instances; polychrest instances; magical instances. Now the use of these instances,
wherein they excel common instances, is found either in the
informative part or in the operative, or in both. As regards
the informative, they assist either the senses or the understanding: the senses, as the five instances of the lamp; the understanding, either by hastening the exclusion of the form, as
solitary instances; or by narrowing and indicating more nearly
the affirmative of the form, as instances migratory, striking, of
companionship, and subjunctive; or by exalting the understanding and leading it to genera and common natures, either
immediately, as instances clandestine, singular, and of alliance, or in the next degree, as constitutive, or in the lowest,
as conformable; or by setting the understanding right when
led astray by habit, as deviating instances; or by leading it to
the great form or fabric of the universe, as bordering instances; or by guarding it against false forms and causes, as

instances of the fingerpost and of divorce. In the operative part they either point out, or measure, or facilitate practice. They point it out by showing with what we should begin, that we may not go again over old ground, as instances of power; or to what we should aspire if means be given, as intimating instances. The four mathematical instances measure practice: polychrest and magical instances facilitate it.

Again, out of these twenty-seven instances there are some of which we must make a collection at once, as I said above, without waiting for the particular investigation of natures. Of this sort are instances conformable, singular, deviating, bordering, of power, of the dose, intimating, polychrest, and magical. For these either help and set right the understanding and senses, or furnish practice with her tools in a general way. The rest need not be inquired into till we come to make Tables of Presentation for the work of the interpreter concerning some particular nature. For the instances marked and endowed with these prerogatives are as a soul amid the common instances of presentation and, as I said at first, a few of them do instead of many; and therefore in the formation of the Tables they must be investigated with all zeal and set down therein. It was necessary to handle them beforehand because I shall have to speak of them in what follows. But now I must proceed to the supports and rectifications of induction, and then to concretes, and Latent Processes, and Latent Configurations, and the rest, as set forth in order in the twenty-first Aphorism; that at length (like an honest and faithful guardian) I may hand over to men their fortunes, now their understanding is emancipated and come as it were of age; whence there cannot but follow an improvement in man's estate and an enlargement of his power over nature. For man by the fall fell at the same time from his state of innocency and from his dominion over creation. Both of these losses however can even in this life be in some part repaired; the former by religion and faith, the latter by arts and sciences. For creation was not by the curse made altogether and forever a rebel, but in virtue of that charter "In the sweat of thy face shalt thou eat bread," it is now by various labors (not certainly by disputations or idle magical ceremonies, but by various labors) at length and in some measure subdued to the supplying of man with bread, that is, to the uses of human life.

PREPARATIVE TOWARD

NATURAL AND EXPERIMENTAL HISTORY

DESCRIPTION OF A NATURAL AND EXPERIMENTAL HISTORY

SUCH AS MAY SERVE FOR THE FOUNDATION
OF A TRUE PHILOSOPHY

My object in publishing my *Instauration* by parts is that some portion of it may be put out of peril. A similar reason induces me to subjoin here another small portion of the work, and to publish it along with that which has just been set forth. This is the description and delineation of a natural and experimental history, such as may serve to build philosophy upon, and containing material true and copious and aptly digested for the work of the interpreter which follows. The proper place for it would be when I come in due course to the "Preparatives" of Inquiry. I have thought it better, however, to introduce it at once without waiting for that. For a history of this kind, such as I conceive and shall presently describe, is a thing of very great size and cannot be executed without great labor and expense, requiring as it does many people to help, and being (as I have said elsewhere) a kind of royal work. It occurs to me, therefore, that it may not be amiss to try if there be any others who will take these matters in hand, so that while I go on with the completion of my original design, this part which is so manifold and laborious may even during my life (if it so please the Divine Majesty) be prepared and set forth, others applying themselves diligently to it along with me; the rather because my own strength (if I should have no one to help me) is hardly equal to such a province. For as much relates to the work itself of the intellect, I shall perhaps be able to master that by myself; but the materials on which the intellect has to work are so widely spread that one must employ factors and merchants to go everywhere in search of them and bring them in. Besides I hold it to be somewhat beneath the dignity of an undertaking like mine that I should spend my own time in a matter which is open to almost every man's industry. That, however, which is the main part of the matter I will myself now supply, by diligently and exactly setting forth the method and description of a history of this kind, such as shall satisfy my intention; lest men for want of warning set to work the wrong way and guide themselves by the example of the natural histories now in use, and so go far astray from my design. Meanwhile, what I have often said I must here emphatically repeat: that if all the wits of all the ages had met or shall hereafter meet together, if the whole human race had applied or shall hereafter apply themselves to philosophy, and the whole earth had been or shall be nothing but academies and colleges and schools of learned men, still without a natural and experimental history such as I am going to prescribe, no progress worthy of the human race could have been made or can be made in philosophy and the sciences. Whereas, on the other hand, let such a history be once provided and well set forth, and let there be added to it such auxiliary and light-giving experiments as in the very course of interpretation will present themselves or will have to be found out, and the investigation of nature and of all sciences will be the work of a few years. This, therefore, must be done or the business must be given up. For in this way, and in this way only, can the foundations of a true and active philosophy be established; and then will men wake as from deep sleep, and at once perceive what a difference there is between the dogmas and figments of the wit and a true and active philosophy, and what it is in questions of nature to consult nature herself.

First, then, I will give general precepts for the composition of this history; then I will set out the particular figure of it, inserting sometimes as well the purpose to which the inquiry is to be adapted and referred as the particular point to be inquired in order that a good understanding and forecast of the mark aimed at may suggest to men's minds other things also which I may perhaps have overlooked. This history I call "Primary History," or the "Mother History."

APHORISMS ON THE COMPOSITION OF THE PRIMARY HISTORY

APHORISM

I

Nature exists in three states, and is subject, as it were, to three kinds of regimen. Either she is free and develops herself in her own ordinary course, or she is forced out of her proper state by the perverseness and insubordination of matter and the violence of impediments, or she is constrained and molded by art and human ministry. The first state refers to the "species" of things; the second to "monsters"; the third to "things artificial." For in things artificial nature takes orders from man and works under his authority; without man, such things would never have been made. But by the help and ministry of man a new face of bodies, another universe or theater of things, comes into view. Natural history therefore is threefold. It treats of the "liberty" of nature, or the "errors" of nature, or the "bonds" of nature, so that we may fairly distribute it into history of "generations," of "pretergenerations," and of "arts"; which last I also call "mechanical" or "experimental" history. And yet I do not make it a rule that these three should be kept apart and separately treated. For why should not the history of the monsters in the several species be joined with the history of the species themselves? And things artificial again may sometimes be rightly joined with the species, though sometimes they will be better kept separate. It will be best, therefore, to consider these things as the case arises. For too much method produces iterations and prolixity as well as none at all.

II

Natural history, which in its subject (as I said) is threefold, is in its use twofold. For it is used either for the sake of the knowledge of the particular things which it contains or as the primary material of philosophy and the stuff and subject matter of true induction. And it is this latter which is now in hand—now, I say, for the first time; nor has it ever been taken in hand till now. For neither Aristotle, nor Theophrastus, nor Dioscorides, nor Caius Plinius ever set this before them as the end of natural history. And the chief part of the matter rests in this, that they who shall hereafter take it upon them to write natural history should bear this continually in mind— that they ought not to consult the pleasure of the reader, no, nor even that utility which may be derived immediately from their narrations, but to seek out and gather together such store and variety of things as may suffice for the formation of true axioms. Let them but remember this, and they will find out for themselves the method in which the history should be composed. For the end rules the method.

III

But the more difficult and laborious the work is, the more ought it to be discharged of matters superfluous. And therefore there are three things upon which men should be warned to be sparing of their labor, as those which will immensely increase the mass of the work and add little or nothing to its worth.

First then, away with antiquities, and citations or testimonies of authors, and also with disputes and controversies and differing opinions—everything, in short, which is philological. Never cite an author except in a matter of doubtful credit; never introduce a controversy unless in a matter of great moment. And for all that concerns ornaments of speech, similitudes, treasury of eloquence, and such like emptinesses, let it be utterly dismissed. Also let all those things which are admitted be themselves set down briefly and concisely, so that they may be nothing less than words. For no man who is collecting and storing up materials for ship building or the like, thinks of arranging them elegantly, as in a shop, and displaying them so as to please the eye; all his care is that they be sound and good, and that they be so arranged as to take up as little room as possible in the warehouse. And this is exactly what should be done here.

Secondly, that superfluity of natural histories in descriptions and pictures of species, and the curious variety of the same, is not much to the purpose. For small varieties of this kind are only a kind of sports and wanton freaks of nature and come near to the nature of individuals. They afford a pleasant recreation in wandering among them and looking at them as objects in themselves, but the information they yield to the sciences is slight and almost superfluous.

Thirdly, all superstitious stories (I do not say stories of prodigies, when the report appears to be faithful and probable, but superstitious stories) and experiments of ceremonial magic should be altogether rejected. For I would not have the infancy of philosophy, to which natural history is as a nursing mother, accustomed to old wives' fables. The time will perhaps come (after we have gone somewhat deeper into the investigation of nature) for a light review of things of this kind, that if there remain any grains of natural virtue in these dregs, they may be extracted and laid up for use. In the meantime they should be set aside. Even the experiments of natural magic should be sifted diligently and severely before they are received, especially those which are commonly derived from vulgar sympathies and antipathies, with great sloth and facility both of believing and inventing.

And it is no small thing to relieve natural history from the three superfluities above mentioned, which would otherwise fill volumes. Nor is this all. For in a great work it is no less necessary that what is admitted should be written succinctly than that what is superfluous should be rejected, though no doubt this kind of chastity and brevity will give less pleasure both to the reader and the writer. But it is always to be remembered that this which we are now about is only a granary and storehouse of matters, not meant to be pleasant to stay or live in, but only to be entered as occasion requires, when anything is wanted for the work of the interpreter which follows.

IV

In the history which I require and design, special care is to
be taken that it be of wide range and made to the measure of
the universe. For the world is not to be narrowed till it will
go into the understanding (which has been done hitherto), but
the understanding to be expanded and opened till it can take
in the image of the world as it is in fact. For that fashion of
taking few things into account, and pronouncing with reference to a few things, has been the ruin of everything. To re-
sume then the divisions of natural history which I made just
now—viz., that it is a history of generations, Pretergenerations,
and arts—I divide the history of generations into five parts.
The first, of ether and things celestial. The second, of meteors
and the regions (as they call them) of air, viz., of the tracts
which lie between the moon and the surface of the earth; to
which part also (for order's sake, however the truth of the
thing may be) I assign comets of whatever kind, both higher
and lower. The third, of earth and sea. The fourth, of the
elements (as they call them), flame or fire, air, water, earth,
understanding, however, by elements, not the first principles of
things, but the greater masses of natural bodies. For the nature
of things is so distributed that the quantity or mass of some
bodies in the universe is very great, because their configurations require a texture of matter easy and obvious, such as are
those four bodies which I have mentioned; while of certain
other bodies the quantity is small and weakly supplied, because the texture of matter which they require is very complex and subtle, and for the most part determinate and organic, such as are the species of natural things—metals, plants,
animals. Hence I call the former kind of bodies the "greater
colleges," the latter the "lesser colleges." Now the fourth part
of the history is of those greater colleges—under the name of
elements, as I said. And let it not be thought that I confound
this fourth part with the second and third, because in each of
them I have mentioned air, water, and earth. For the history
of these enters into the second and third, as they are integral

parts of the world, and as they relate to the fabric and configuration of the universe. But in the fourth is contained the
history of their own substance and nature, as it exists in their
several parts of uniform structure, and without reference to
the whole. Lastly, the fifth part of the history contains the
lesser colleges, or species, upon which natural history has
hitherto been principally employed.

As for the history of pretergenerations, I have already said
that it may be most conveniently joined with the history of
generations—I mean the history of prodigies which are natural.
For the superstitious history of marvels (of whatever kind) I
remit to a quite separate treatise of its own; which treatise I do
not wish to be undertaken now at first, but a little after, when
the investigation of nature has been carried deeper.

History of arts, and of nature as changed and altered by
man, or experimental history, I divide into three. For it is
drawn either from mechanical arts, or from the operative part
of the liberal arts, or from a number of crafts and experiments
which have not yet grown into an art properly so called, and
which sometimes indeed turn up in the course of most ordinary experience and do not stand at all in need of art.

As soon, therefore, as a history has been completed of all
these things which I have mentioned—namely, generations,
pretergenerations, arts, and experiments, it seems that nothing
will remain unprovided whereby the sense can be equipped for
information of the understanding. And then shall we be no
longer kept dancing within little rings, like persons bewitched,
but our range and circuit will be as wide as the compass of
the world.

V

Among the parts of history which I have mentioned, the
history of arts is of most use because it exhibits things in motion and leads more directly to practice. Moreover, it takes off
the mask and veil from natural objects, which are commonly
concealed and obscured under the variety of shapes and external appearance. Finally, the vexations of art are certainly

as the bonds and handcuffs of Proteus, which betray the ultimate struggles and efforts of matter. For bodies will not be
destroyed or annihilated, rather than that they will turn themselves into various forms. Upon this history, therefore, mechanical and illiberal as it may seem (all fineness and daintiness
set aside), the greatest diligence must be bestowed.

Again, among the particular arts those are to be preferred
which exhibit, alter, and prepare natural bodies and materials
of things, such as agriculture, cookery, chemistry, dyeing, the
manufacture of glass, enamel, sugar, gunpowder, artificial fires,
paper, and the like. Those which consist principally in the
subtle motion of the hands or instruments are of less use, such
as weaving, carpentry, architecture, manufacture of mills,
clocks, and the like, although these too are by no means to be
neglected, both because many things occur in them which relate to the alterations of natural bodies, and because they give
accurate information concerning local motion, which is a thing
of great importance in very many respects.

But in the whole collection of this history of arts it is especially to be observed and constantly borne in mind that not
only those experiments in each art which serve the purpose of
the art itself are to be received, but likewise those which turn
up anyhow by the way. For example, that locusts or crabs,
which were before of the color of mud, turn red when baked
is nothing to the table; but this very instance is not a bad one
for investigating the nature of redness, seeing that the same
thing happens in baked bricks. In like manner the fact that
meat is sooner salted in winter than in summer is not only
important for the cook that he may know how to regulate the
pickling, but is likewise a good instance for showing the
nature and impression of cold. Therefore, it would be an
utter mistake to suppose that my intention would be satisfied
by a collection of experiments of arts made only with the view
of thereby bringing the several arts to greater perfection. For
though this be an object which in many cases I do not despise,
yet my meaning plainly is that all mechanical experiments
should be as streams flowing from all sides into the sea of

philosophy. But how to select the more important instances in every kind (which are principally and with the greatest diligence to be sought and as it were hunted out) is a point to be learned from the prerogatives of instances.

VI

In this place also is to be resumed that which in the 99th, 119th, and 120th aphorisms of the first book I treated more at large, but which it may be enough here to enjoin shortly by way of precept—namely, that there are to be received into this history, first, things the most ordinary, such as it might be thought superfluous to record in writing because they are so familiarly known; secondly, things mean, illiberal, filthy (for "to the pure all things are pure," and if money obtained from Vespasian's tax smelt well, much more does light and information from whatever source derived); thirdly, things trifling and childish (and no wonder, for we are to become again as little children); and lastly, things which seem over subtle, because they are in themselves of no use. For the things which will be set forth in this history are not collected (as I have already said) on their own account; and therefore neither is their importance to be measured by what they are worth in themselves, but according to their indirect bearing upon other things and the influence they may have upon philosophy.

VII

Another precept is that everything relating both to bodies and virtues in nature be set forth (as far as may be) numbered, weighed, measured, defined. For it is works we are in pursuit of, not speculations; and practical working comes of the due combination of physics and mathematics. And therefore the exact revolutions and distances of the planets—in the history of the heavenly bodies; the compass of the land and the superficial space it occupies in comparison of the waters—in the history of earth and sea; how much compression air will bear without strong resistance—in the history of air; how much one

metal outweighs another—in the history of metals; and numberless other particulars of that kind are to be ascertained and
set down. And when exact proportions cannot be obtained,
then we must have recourse to indefinite estimates and com-
paratives. As for instance (if we happen to distrust the calculations of astronomers as to the distances of the planets), that
the moon is within the shadow of the earth, that Mercury is
beyond the moon, and the like. Also when mean proportions
cannot be had, let extremes be proposed, as that a weak magnet will raise so many times its own weight of iron, while the
most powerful will raise sixty times its own weight (as I have
myself seen in the case of a very small armed magnet). I know
well enough that these definite instances do not occur readily
or often, but that they must be sought for as auxiliaries in the
course of interpretation itself when they are most wanted. But
nevertheless if they present themselves accidentally, provided
they do not too much interrupt the progress of the natural
history, they should also be entered therein.

VIII

With regard to the credit of the things which are to be
admitted into the history, they must needs be either certainly
true, doubtful whether true or not, or certainly not true.
Things of the first kind should be set down simply; things of
the second kind with a qualifying note, such as "it is reported,"
"they relate," "I have heard from a person of credit," and the
like. For to add the arguments on either side would be too
laborious and would certainly interrupt the writer too much.
Nor is it of much consequence to the business in hand because
(as I have said in the 118th aphorism of the first book) mistakes in experimenting, unless they abound everywhere, will
be presently detected and corrected by the truth of axioms.
And yet if the instance be of importance, either from its own
use or because many other things may depend upon it, then
certainly the name of the author should be given, and not the
name merely, but it should be mentioned withal whether he

took it from report, oral or written (as most of Pliny's statements are), or rather affirmed it of his own knowledge; also whether it was a thing which happened in his own time or earlier; and again, whether it was a thing of which, if it really happened, there must needs have been many witnesses; and finally, whether the author was a vain-speaking and light person or sober and severe; and the like points, which bear upon the weight of the evidence. Lastly, things which though certainly not true are yet current and much in men's mouths, having either through neglect or from the use of them in similitudes prevailed now for many ages (as that the diamond binds the magnet, garlic weakens it, that amber attracts everything except basil, and other things of that kind), these it will not be enough to reject silently; they must be in express words proscribed, that the sciences may be no more troubled with them.

Besides, it will not be amiss, when the source of any vanity or credulity happens to present itself, to make a note of it, as, for example, that the power of exciting Venus is ascribed to the herb Satyrion because its root takes the shape of testicles— when the real cause of this is that a fresh bulbous root grows upon it every year, last year's root still remaining; whence those twin bulbs. And it is manifest that this is so, because the new root is always found to be solid and succulent, the old withered and spongy. And therefore it is no marvel if one sinks in water and the other swims—which nevertheless goes for a wonder and has added credit to the other virtues ascribed to this herb.

IX

There are also some things which may be usefully added to the natural history, and which will make it fitter and more convenient for the work of the interpreter, which follows. They are five.

First, questions (I do not mean as to causes but as to the fact) should be added in order to provoke and stimulate further inquiry, as in the history of earth and sea, whether the Caspian ebbs and flows, and at how many hours' interval; whether there is any southern continent or only islands, and the like.

Secondly, in any new and more subtle experiment the manner in which the experiment was conducted should be added, that men may be free to judge for themselves whether the information obtained from that experiment be trustworthy or fallacious, and also that men's industry may be roused to discover, if possible, methods more exact.

Thirdly, if in any statement there be anything doubtful or questionable, I would by no means have it suppressed or passed in silence, but plainly and perspicuously set down by way of note or admonition. For I want this primary history to be compiled with a most religious care, as if every particular were stated upon oath, seeing that it is the book of God's works and (so far as the majesty of heavenly may be compared with the humbleness of earthly things) a kind of second Scripture.

Fourthly, it would not be amiss to intersperse observations occasionally, as Pliny has done; as in the history of earth and sea, that the figure of the earth (as far as it is yet known) compared with the seas is narrow and pointed toward the south, wide and broad toward the north, the figure of the sea contrary; that the great oceans intersect the earth in channels running north and south, not east and west, except perhaps in the extreme polar regions. It is also very good to add canons (which are nothing more than certain general and catholic observations), as in the history of the heavenly bodies, that Venus is never distant more than 46 parts from the sun, Mercury never more than 23, and that the planets which are placed above the sun move slowest when they are furthest from the earth, those under the sun fastest. Moreover, there is another kind of observation to be employed, which has not yet come into use, though it be of no small importance. This is, that to the enumeration of things which are should be subjoined an enumeration of things which are not. As in the history of the heavenly bodies, that there is not found any star oblong or triangular, but that every star is globular—either globular simply, as the moon, or apparently angular, but globular in the middle, as the other stars, or apparently radiant but globular in the middle, as the sun—or that the stars are scattered about the sky in no order at all, so that there is not found among them either quincunx or square, or any other regular figure (howsoever the names be given of Delta, Crown, Cross, Chariot, etc.) scarcely so much as a straight line, except perhaps in the belt and dagger of Orion.

Fifthly, that may perhaps be of some assistance to an inquirer which is the ruin and destruction of a believer; viz., a brief review, as in passage, of the opinions now received, with their varieties and sects, that they may touch and rouse the intellect and no more.

X

And this will be enough in the way of general precepts which, if they be diligently observed, the work of the history will at once go straight toward its object and be prevented from increasing beyond bounds. But if even as here circumscribed and limited it should appear to some poor-spirited person a vast work—let him turn to the libraries; and there among other things let him look at the bodies of civil and canonical law on one side, and at the commentaries of doctors and lawyers on the other, and see what a difference there is between the two in point of mass and volume. For we (who as faithful secretaries do but enter and set down the laws themselves of nature and nothing else) are content with brevity, and almost compelled to it by the condition of things; whereas opinions, doctrines, and speculations are without number and without end.

And whereas in the Plan of the Work I have spoken of the *cardinal virtues* in nature, and said that a history of these must also be collected and written before we come to the work of Interpretation, I have not forgotten this, but I reserve this part for myself since until men have begun to be somewhat more closely intimate with nature, I cannot venture to rely very much on other people's industry in that matter.

And now should come the delineation of the particular histories. But I have at present so many other things to do that I can only find time to subjoin a Catalogue of their titles. As soon, however, as I have leisure for it, I mean to draw up a set of questions on the several subjects, and to explain what points with regard to each of the histories are especially to be inquired and collected, as conducing to the end I have in view—like a kind of particular topics. In other words, I mean (according to the practice in civil causes) in this great plea or suit granted by the divine favor and providence (whereby the human race seeks to recover its right over nature), to examine nature herself and the arts upon interrogatories.

CATALOGUE OF PARTICULAR HISTORIES
BY TITLES

1. History of the Heavenly Bodies; or Astronomical History.
2. History of the Configuration of the Heaven and the parts thereof toward the Earth and the parts thereof; or Cosmographical History.
3. History of Comets.
4. History of Fiery Meteors.
5. History of Lightnings, Thunderbolts, Thunders, and Coruscations.
6. History of Winds and Sudden Blasts and Undulations of the Air.
7. History of Rainbows.
8. History of Clouds, as they are seen above.
9. History of the Blue Expanse, of Twilight, of Mock-Suns, Mock-Moons, Haloes, various colors of the Sun; and of every variety in the aspect of the heavens caused by the medium.
10. History of Showers, Ordinary, Stormy, and Prodigious; also of Waterspouts (as they are called); and the like.
11. History of Hail, Snow, Frost, Hoar-frost, Fog, Dew, and the like.
12. History of all other things that fall or descend from above, and that are generated in the upper region.
13. History of Sounds in the upper region (if there be any), besides Thunder.
14. History of Air as a whole, or in the Configuration of the World.
15. History of the Seasons or Temperatures of the Year, as well according to the variations of Regions as according to accidents of Times and periods of Years; of Floods, Heats, Droughts, and the like.

285

16. History of Earth and Sea; of the Shape and Compass of them, and their Configurations compared with each other; and of their broadening or narrowing; of Islands in the Sea; of Gulfs of the Sea, and Salt Lakes within the Land; Isthmuses and Promontories.

17. History of the Motions (if any be) of the Globe of Earth and Sea; and of the Experiments from which such motions may be collected.

18. History of the greater Motions and Perturbations in Earth and Sea; Earthquakes, Tremblings and Yawnings of the Earth, Islands newly appearing; Floating Islands; Breakings off of Land by entrance of the Sea, Encroachments and Inundations and contrariwise Recessions of the Sea; Eruptions of Fire from the Earth; Sudden Eruptions of Waters from the Earth; and the like.

19. Natural History of Geography; of Mountains, Valleys, Woods, Plains, Sands, Marshes, Lakes, Rivers, Torrents, Springs, and every variety of their course, and the like; leaving apart Nations, Provinces, Cities, and such like matters pertaining to Civil life.

20. History of Ebbs and Flows of the Sea; Currents, Undulations, and other Motions of the Sea.

21. History of the other Accidents of the Sea; its Saltness, its various Colors, its Depth; also of Rocks, Mountains and Valleys under the Sea, and the like.

Next come Histories of the Greater Masses.

22. History of Flame and of things Ignited.
23. History of Air, in Substance, not in the Configuration of the World.
24. History of Water, in Substance, not in the Configuration of the World.
25. History of Earth and the diversity thereof, in Substance, not in the Configuration of the World.

Next come Histories of Species.

26. History of perfect Metals, Gold, Silver; and of the Mines, Veins, Marcasites of the same; also of the Working in the Mines.
27. History of Quicksilver.
28. History of Fossils; as Vitriol, Sulphur, etc.
29. History of Gems; as the Diamond, the Ruby, etc.
30. History of Stones; as Marble, Touchstone, Flint, etc.
31. History of the Magnet.
32. History of Miscellaneous Bodies, which are neither entirely Fossil nor Vegetable; as Salts, Amber, Ambergris, etc.
33. Chemical History of Metals and Minerals.
34. History of Plants, Trees, Shrubs, Herbs; and of their parts, Roots, Stalks, Wood, Leaves, Flowers, Fruits, Seeds, Gums, etc.
35. Chemical History of Vegetables.
36. History of Fishes, and the Parts and Generation of them.
37. History of Birds, and the Parts and Generation of them.
38. History of Quadrupeds, and the Parts and Generation of them.
39. History of Serpents, Worms, Flies, and other insects; and of the Parts and Generation of them.
40. Chemical History of the things which are taken by Animals.

Next come Histories of Man.

41. History of the Figure and External Limbs of Man, his Stature, Frame, Countenance and Features; and of the variety of the same according to Races and Climates, or other smaller differences.
42. Physiognomical History of the same.
43. Anatomical History, or of the Internal Members of Man; and of the variety of them, as it is found in the

Natural Frame and Structure, and not merely as regards Diseases and Accidents out of the course of Nature.

44. History of the parts of Uniform Structure in Man; as Flesh, Bones, Membranes, etc.

45. History of Humors in Man; Blood, Bile, Seed, etc.

46. History of Excrements; Spittle, Urine, Sweats, Stools, Hair of the Head, Hairs of the Body, Whitlows, Nails, and the like.

47. History of Faculties; Attraction, Digestion, Retention, Expulsion, Sanguification, Assimilation of Aliment into the members, conversion of Blood and Flower of Blood into Spirit, etc.

48. History of Natural and Involuntary Motions; as Motion of the Heart, the Pulses, Sneezing, Lungs, Erection, etc.

49. History of Motions partly Natural and partly Violent; as of Respiration, Cough, Urine, Stool, etc.

50. History of Voluntary Motions; as of the Instruments of Articulation of Words; Motions of the Eyes, Tongue, Jaws, Hands, Fingers; of Swallowing, etc.

51. History of Sleep and Dreams.

52. History of different habits of Body—Fat, Lean; of the Complexions (as they call them), etc.

53. History of the Generation of Man.

54. History of Conception, Vivification, Gestation in the Womb, Birth, etc.

55. History of the Food of Man; and of all things Eatable and Drinkable; and of all Diet; and of the variety of the same according to nations and smaller differences.

56. History of the Growth and Increase of the Body, in the whole and in its parts.

57. History of the Course of Age; Infancy, Boyhood, Youth, Old Age; of Length and Shortness of Life, and the like, according to nations and lesser differences.

58. History of Life and Death.

59. History Medicinal of Diseases, and the Symptoms and Signs of them.
60. History Medicinal of the Treatment and Remedies and Cures of Diseases.
61. History Medicinal of those things which preserve the Body and the Health.
62. History Medicinal of those things which relate to the Form and Comeliness of the Body.
63. History Medicinal of those things which alter the Body, and pertain to Alternative Regimen.
64. History of Drugs.
65. History of Surgery.
66. Chemical History of Medicines.
67. History of Vision, and of things Visible.
68. History of Painting, Sculpture, Modelling, etc.
69. History of Hearing and Sound.
70. History of Music.
71. History of Smell and Smells.
72. History of Taste and Tastes.
73. History of Touch, and the objects of Touch.
74. History of Venus, as a species of Touch.
75. History of Bodily Pains, as species of Touch.
76. History of Pleasure and Pain in general.
77. History of the Affections; as Anger, Love, Shame, etc.
78. History of the Intellectual Faculties; Reflection, Imagination, Discourse, Memory, etc.
79. History of Natural Divinations.
80. History of Diagnostics, or Secret Natural Judgments.

81. History of Cookery, and the arts thereto belonging, as of the Butcher, Poulterer, etc.
82. History of Baking, and the Making of Bread, and the arts thereto belonging, as of the Miller, etc.
83. History of Wine.
84. History of the Cellar and of different kinds of Drink.
85. History of Sweetmeats and Confections.

86. History of Honey.
87. History of Sugar.
88. History of the Dairy.
89. History of Baths and Ointments.
90. Miscellaneous History concerning the care of the body —as of Barbers, Perfumers, etc.
91. History of the working of Gold, and the arts thereto belonging.
92. History of the manufactures of Wool, and the arts thereto belonging.
93. History of the manufactures of Silk, and the arts thereto belonging.
94. History of manufactures of Flax, Hemp, Cotton, Hair, and other kinds of Thread, and the arts thereto belonging.
95. History of manufactures of Feathers.
96. History of Weaving, and the arts thereto belonging.
97. History of Dyeing.
98. History of Leather-making, Tanning, and the arts thereto belonging.
99. History of Ticking and Feathers.
100. History of working in Iron.
101. History of Stone-cutting.
102. History of the making of Bricks and Tiles.
103. History of Pottery.
104. History of Cements, etc.
105. History of working in Wood.
106. History of working in Lead.
107. History of Glass and all vitreous substances, and of Glass-making.
108. History of Architecture generally.
109. History of Wagons, Chariots, Litters, etc.
110. History of Printing, of Books, of Writing, of Sealing; of Ink, Pen, Paper, Parchment, etc.
111. History of Wax.
112. History of Basket-making.

113. History of Mat-making, and of manufactures of Straw, Rushes, and the like.
114. History of Washing, Scouring, etc.

115. History of Agriculture, Pasturage, Culture of Woods, etc.
116. History of Gardening.
117. History of Fishing.
118. History of Hunting and Fowling.
119. History of the Art of War, and of the arts thereto belonging, as Armory, Bow-making, Arrow-making, Musketry, Ordnance, Cross-bows, Machines, etc.
120. History of the Art of Navigation, and of the crafts and arts thereto belonging.
121. History of Athletics and Human Exercises of all kinds.
122. History of Horsemanship.
123. History of Games of all kinds.
124. History of Jugglers and Mountebanks.
125. Miscellaneous History of various Artificial Materials,— as Enamel, Porcelain, various Cements, etc.
126. History of Salts.
127. Miscellaneous History of various Machines and Motions.
128. Miscellaneous History of Common Experiments which have not grown into an Art.

Histories must also be written of Pure Mathematics; though they are rather observations than experiments.

129. History of the Natures and Powers of Numbers.
130. History of the Natures and Powers of Figures.

It may not be amiss to observe that, whereas many of the experiments must come under more titles than one (as the history of plants and the history of the art of gardening have many things in common), it will be more convenient to investigate them with reference to arts, and to arrange them with reference to bodies. For I care little about the mechanical arts themselves: only about those things which they contribute to the equipment of philosophy. But these things will be better regulated as the case arises.